GENES
EMBRYOS

Frontiers in Molecular Biology

Series editors

B.D.Hames

Department of Biochemistry, University of Leeds, Leeds LS2 9JT, UK

D.M.Glover

Cancer Research Campaign, Eukaryotic Molecular Genetics
Research Group, Department of Biochemistry, Imperial College
of Science and Technology, London SW7 2AZ, UK

Other titles in the series:

Gene Rearrangements

Molecular Immunology

Molecular Neurobiology

Oncogenes

Transcription and Splicing

GENES AND EMBRYOS

Edited by

D.M.Glover

Cancer Research Campaign, Eukaryotic Molecular Genetics
Research Group, Department of Biochemistry, Imperial College
of Science and Technology, London SW7 2AZ, UK

and

B.D.Hames

Department of Biochemistry, University of Leeds, Leeds LS2 9JT,
UK

IRL PRESS
—at—
OXFORD UNIVERSITY PRESS
Oxford New York Tokyo

IRL Press
Eynsham
Oxford
England

First Published 1989

British Library Cataloguing in Publication Data

Genes and embryos
1. Animals. Embryos. Genes
I. Glover, David M. II. Hames, B.D. (B. David)
III. Series
591.87'322

Library of Congress Cataloging in Publication Data

Genes and embryos/edited by D.M. Glover and B.D. Hames
p. cm. -- (Frontiers in molecular biology)
Includes bibliographies and index.
ISBN 0-19-963028-3: $54.00. -- ISBN 0-19-963029-1 (pbk.): $36.00
1. Embryology, 2. Genetic regulation. 3. Molecular biology.
I. Glover, David M. II. Hames, B.D. III. Series.
QL955.G36 1989
591.3'3--dc20

ISBN 0 19 963028 3 (hardbound)
ISBN 0 19 963029 1 (softbound)

Previously announced as:

ISBN 1 85221 150 4 (hardbound)
ISBN 1 85221 151 2 (softbound)

Typeset and printed by Information Press Ltd, Oxford, England.

Preface

This book reviews the exciting discoveries made over recent years which contribute to our understanding of development at a molecular level. It is limited to a discussion of four organisms that have provided model systems for the study of development since the turn of the century. Models proposed early in this century from work with amphibians still form the basis of several active areas of research described by Thomas Sargent in his chapter on *Xenopus* development. These studies are now greatly facilitated by the availability of molecular probes. Studies on *Drosophila* genetics stretch back just as far, but the genetics of the embryonic development of the fly have exploded over the past ten years. So much information has accumulated in this area that we have asked two sets of contributors to assess the field. Kathryn Anderson's chapter looks at the maternal contribution to the development of the *Drosophila* embryo, and Michael Levine and Kate Harding have considered the contribution of genes active in the zygote itself. Genetics plays an equally important role in the studies of the early development of *Caenorhabditis*, described by Kenneth Kemphues. Genes that play a key role in development, originally discovered in *Drosophila*, have analogs in many organisms including vertebrates. These are encountered both in Thomas Sargent's chapter, and in Ian Jackson's chapter on the early development of the mouse. Once again, it is clear from work with the mouse how a genetic approach is proceeding hand in hand with studies that use both cell and molecular biology. Molecular biology has been defined as 'that which interests molecular biologists'. This is certainly true in the area of development, and we hope that in this volume it is possible to have a glimpse of the multidisciplinary approaches that are being used to answer some of the most fundamental questions in biology.

David Glover
David Hames

Contributors

Kathryn V.Anderson
Department of Molecular Biology, University of California, Berkeley, CA 94720, USA

Katherine W.Harding
Department of Biological Sciences, Fairchild Center, Columbia University, NY 10027, USA

Ian J.Jackson
Developmental Genetics Group, MRC Human Genetics Unit, Western General Hospital, Crewe Road, Edinburgh EH4 2XU, UK

Kenneth J.Kemphues
Section of Genetics and Development, Cornell University, 142 Emerson Hall, Ithaca, NY 14853, USA

Michael S.Levine
Department of Biological Sciences, Fairchild Center, Columbia University, NY 10027, USA

Thomas D.Sargent
Laboratory of Molecular Genetics, National Institute of Child Health and Human Development, National Institutes of Health, Bethesda, MD 20892, USA

Contents

Abbreviations

abd-A	*abdominal-A*
Abd-B	*Abdominal-B*
Antp	*Antennapedia*
ANTP-C	*Antennapedia* complex
A/P	anterior – posterior
bcd	*bicoid*
BRL	buffalo rat liver
BX-C	*Bithorax* complex
CAT	chloramphenicol acetyl transferase
cad	*caudal*
CNS	central nervous system
Dfd	*Deformed*
dpc	days *post coitum*
DT	diphtheria toxin
EGF	epithelial growth factor
en	*engrailed*
EC	embryonal carcinoma
ES	embryonal stem
esc	*extra sex combs*
FGF	fibroblast growth factor
ftz	*fushi tarazu*
GF	growth factor
gt	*giant*
h	*hairy*
Hox	*homeobox*
hb	*hunchback*
HPRT	hypoxanthine phosphoribosyl transferase
HSV-TK	herpes simplex virus – thymidine kinase
ICM	inner cell mass
IGF	insulin-like growth factor
kni	*knirps*
Kr	*Krüppel*
MBT	midblastula transition
MIS	Mullerian inhibiting substance
NCAM	neural cell adhesion molecule

odd	*odd-skipped*
opa	*odd-paired*
Pc	*Polycomb*
PCR	polymerase chain reaction
PDGF	platelet-derived growth factor
PGC	primordial germ cell
prd	*paired*
PS	parasegment
run	*runt*
ptc	*patched*
Scr	*Sex combs reduced*
TGFβ	transforming growth factor-β
tll	*tailless*
Ubx	*Ultrabithorax*
wg	*wingless*
X-Gal	5-bromo-4-chloro-3-indolyl-β-D-galactopyranoside

1

Drosophila: the maternal contribution

Kathryn V.Anderson

1. Introduction

> I would like to ascribe to the cytoplasm of the sea urchin embryo only the initial and simplest properties responsible for differentiation . . . The structure of the egg cytoplasm takes care . . . of the purely 'promorphological' tasks, that is, it provides the most general basic form, the framework within which all specific details are filled in by the nucleus.
>
> *Theodor Boveri, 1902 (1)*

Nearly 90 years ago, Boveri's observations on the importance of chromosomal constitution for the pattern of development of the sea urchin embryo led him to conclude that the role of maternal information in embryonic development is to provide coarse, global cues—the promorphological framework—for embryonic development. As the result of recent genetic studies, it can be concluded that, in *Drosophila* as well, maternally provided components define the basic spatial organization of the embryo, and that the detailed pattern is the result of the activity of zygotic nuclei.

In the past 10 years, systematic mutant searches have identified more than 100 genes that are necessary maternally, in the zygote, or both maternally and zygotically, to allow normal development of the *Drosophila* embryo (2–6, C.Nüsslein-Volhard, G.Jürgens, K.Anderson and R.Lehmann, unpublished results). Maternal-effect mutations have been identified as those female sterile mutations in which eggs are laid, but embryonic development is abnormal. These mutations alter the activity of genes that are transcribed in the female during oogenesis but have effects after fertilization on the development of the embryo. The maternal-effect mutations have defined the specific maternal contribution to embryonic development and provide the basis for the studies that will be described in this chapter.

The characterization of maternal-effect mutants has demonstrated the involvement of maternally encoded gene products in a wide range of embryonic processes, including early nuclear division, nuclear migration, cellularization, gastrulation, definition of the polarity of the body axes, segmentation, neurogenesis and segment identification. Three kinds of tools have been especially useful in characterizing the nature of maternal information: classical genetics, phenotypic rescue by cytoplasmic injection and molecular biology. It is the combination of these three kinds of experiments that has made *Drosophila* developmental genetics so powerful.

Classical genetic analysis of maternal-effect mutations has defined a rather small number of maternally encoded pathways that promote normal embryogenesis and has been successful in assigning specific functions to the products of the individual genes that make up these pathways. The pathways defined by mutant phenotypes reflect the underlying logic of development, which would have been difficult to discern with purely embryological or molecular experiments. The characterization of the phenotypes associated with null and weak alleles, the determination of temperature-sensitive periods of temperature-sensitive alleles, and the analysis of double mutant phenotypes have suggested which wild-type gene products are normally required for specific developmental processes, the order of action of particular gene products and how those products interact with one another.

The embryonic phenotypes caused by many maternal-effect mutants can be partially or completely rescued by injection of wild-type cytoplasm after fertilization. Phenotypic rescue has been an invaluable bioassay for the localization, local requirements, interdependence and molecular nature of these gene products.

A number of the maternally active genes that have been implicated in controlling the pattern of early embryonic development have been cloned. The powerful repertoire of molecular techniques available in *Drosophila* makes it possible to isolate any gene of interest, study the biochemical nature of its product and identify functionally important features of the gene by reintroducing altered versions into the genome (7). Molecular analysis of cloned genes has confirmed many of the conclusions from genetics and rescue experiments, and has opened the door to understanding the biochemical processes that underlie early embryonic development.

This chapter will first provide a general background on the nature of the maternal contribution to embryonic development of *Drosophila*. The majority of the chapter will focus on the studies on the maternally encoded products that organize the spatial pattern of the animal. In the two cases that have been studied in detail, it has been possible to show that the maternal framework for the spatial pattern of embryonic development is based on morphogen gradients.

2. Oogenesis: the construction of the egg

Within the ovary, three different cell types contribute to the production of the egg, the oocyte, the nurse cells and the follicle cells (*Figure 1*) (8).

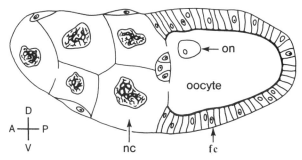

Figure 1.   Cell types in a stage 10 ovarian follicle, a fairly late stage of oogenesis. The polyploid nurse cells (nc) lie anterior to their sister cell, the oocyte. At this stage, the transfer of nurse cell cytoplasm to the oocyte is about half complete. The follicle cell (fc), mesodermal derivatives, surround the oocyte and have begun to secrete the eggshell at the stage shown here. The oocyte nucleus (on) lies in the anterior, dorsal region of the oocyte. D, V, A and P refer to dorsal, ventral, anterior and posterior, respectively.

The germ line cells, set aside very early in development as the pole cells, give rise to the oocyte and its sister cells, the nurse cells. Each oocyte is connected by cytoplasmic bridges to 15 nurse cells. The nurse cells, which undergo several rounds of endoduplication to become approximately 1000-fold polyploid, synthesize most of the cytoplasmic components of the egg. Most of the nurse cell cytoplasm is transferred to the oocyte in the late stage of oogenesis. The oocyte nucleus does not make an important quantitative contribution to the RNA accumulated in the oocyte, although there is a burst of RNA synthesis in the oocyte nucleus at the stage when the nurse cell cytoplasm is transferred to the oocyte. The follicle cells, which surround the nurse cell – oocyte complex, are derived from a totally separate lineage and are mesodermal in origin. The follicle cells are responsible for the construction of the eggshell, the uptake of yolk proteins from the circulating hemolymph, and also play an important role in defining the polarity of the egg. These three cell types not only provide the raw materials necessary to support the 24 h of embryogenesis, they also play an important role in defining the spatial organization of the embryo.

Both genetic and molecular measurements show that a large fraction of the genome is required during oogenesis to support the production of a normal egg. The genetic estimates are based on genetic mosaics in which the somatic cells are wild-type while the germ line cells (oocyte and nurse cells) are homozygous for lethal mutations. Such studies indicate that at least 75% of all the genes that are essential for the viability of the fly are also essential in the germ line to allow survival of the oocyte (9,10). The molecular estimates, based on the total sequence complexity of the RNA found in the ovary, indicate that 10% of all single copy sequences in the genome are represented in ovarian RNA (11). This corresponds to 5000 different mRNA sequences, or 50 – 100% of the number of genes thought to be active during the lifespan of the fly.

Many of the genes necessary for the development of the oocyte must encode products that are necessary for the metabolism of all cells. Since after fertilization the embryo undergoes rapid nuclear divisions in the absence of significant new transcription (12), these genes must be expressed at extraordinarily high levels during oogenesis to provide the maternal dowry that supports the early embryo.

There are also a number of specialized functions required during oogenesis that require high levels of gene expression, including the synthesis of eggshell and yolk uptake. These processes are quite complex. For instance, production of the outer layer of the eggshell, the chorion, requires the synthesis of the structural proteins of the chorion. The synthesis of chorion proteins is regulated at several levels. Chorion mRNAs are produced in a narrow time window at such high levels that the chorion genes must be amplified in the follicle cells (13). In addition, the chorion genes are activated in a precise spatial and temporal sequence in particular subsets of follicle cells to create the asymmetric shape of the eggshell (14).

As in other animals, only a small subset of the genes essential for oogenesis are directly involved in generating the spatial information used in the early embryo, making it difficult to identify those gene products biochemically. Systematic screens for maternal-effect mutations that affect early embryonic development have been carried out for the entire *Drosophila* genome, by testing females homozygous for randomly induced mutations for the ability to produce progeny (3–6, C.Nüsslein-Volhard, G.Jürgens, K.Anderson and R.Lehmann, unpublished results). These screens have found that only 2–5% of all genes are specifically required in the mother to allow normal embryonic development. The genes identified in these screens include those that are required for replication, cellularization and morphogenesis, as well as those that are required for the spatial organization of the body plan. The mutants have provided the basis for studies of the maternal components specifically required for the early development of the animal.

3. Cell biology of the early embryo

After fertilization, embryogenesis proceeds rapidly, with the first instar larva hatching from the eggshell after 24 h (*Figure 2*). Nuclear division cycles occur extremely rapidly, with nine nuclear doublings in the first 90 min of development. These nuclear divisions take place in a syncytium, and are not accompanied by cell division. After the ninth nuclear division, the nuclei migrate from the central yolk region of the egg to the periphery, where they go through four more syncytial divisions in the next hour. After the 13th nuclear division, cell membranes form between the nuclei.

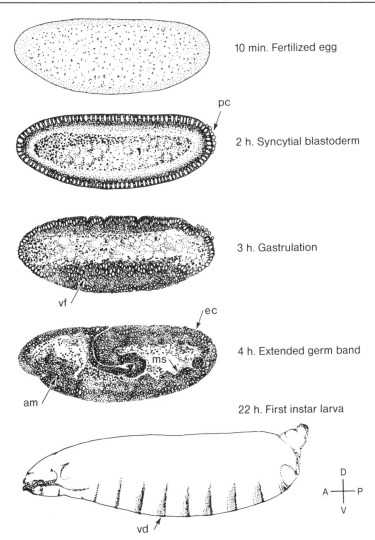

10 min. Fertilized egg

2 h. Syncytial blastoderm

3 h. Gastrulation

4 h. Extended germ band

22 h. First instar larva

Figure 2. Stages of early embryogenesis that rely extensively on maternally encoded products. The figure depicts embryos as they would appear in parasagittal sections. Times refer to development at 25°C. At 10 min after fertilization, the single nucleus lies anteriorly and dorsally in the egg. At the syncytial blastoderm stage, the pole cells (pc) have formed at the posterior end of the embryo. The somatic nuclei are still syncytial and now lie at the periphery of the embryo. As soon as cellularization is completed, gastrulation begins. The first event of gastrulation is the invagination of the cells on the ventral midline in the ventral furrow (vf) to generate the presumptive mesoderm. By 4 h after fertilization, the germ band [ectoderm (ec) plus mesoderm (ms)] has extended along the dorsal side (am, anterior midgut). By this stage, many maternal products have already exerted their effect. At 22 h after fertilization, the fully differentiated first instar larva hatches. The larval cuticle, illustrated here, has a rich spatial pattern, including the segmentally repeated ventral denticle bands (vd). D, V, A and P refer to dorsal, ventral, anterior and posterior, respectively.

As soon as cellularization is complete, at about 3.5 h after fertilization, gastrulation begins. Transcription in zygotic nuclei cannot be detected until the nuclei reach the periphery, and the rate of transcription per nucleus remains low until the end of the syncytial divisions (12). Thus these early stages must rely on maternal products to support the extraordinary high levels of DNA replication and nuclear assembly, as well as nuclear migration and the process of cellularization.

Biochemical studies have shown that histones, tubulin, actin and other molecules that are important for early cleavage are stored at high levels in the egg as either mRNA or protein (15–18). Mutations in these structural genes would certainly cause zygotic lethality, and would not have been isolated as maternal-effect mutations. Eventually it may be possible to use a reverse genetic approach, by mutagenizing cloned structural genes *in vitro* and reintroducing them into the fly genome, to affect the maternal activity of these genes specifically in order to better define their function in the early embryo.

A large fraction of all maternal-effect mutations disrupt the processes of early nuclear division and cellularization. From screens for maternal-effect mutations on the first and second chromosomes, it appears that approximately two-thirds of the maternally active genes that are essential for embryonic development are required for these early steps (6,19). In most cases, the phenotypes associated with these mutations have been analyzed only superficially. A quarter to a third of these early mutants appear unfertilized. Another class undergoes abnormal early replication. In another class, nuclear migration is abnormal, and in 30–40% of the mutants with early visible defects, cellularization is abnormal.

The phenotypes associated with a few mutants have been studied in some detail. In embryos produced by females homozygous for mutant alleles of *giant nuclei (gnu)*, DNA replication is dissociated from nuclear division (20). DNA replication and centrosomal doubling appear to occur normally, but since nuclear division fails to take place, a small number of polyploid nuclei result. Mutations in other loci produce a similar phenotype (19). Three maternal-effect mutants characterized by Rice and Garen (21) affect cellularization of specific regions of the blastoderm. Strikingly, although cellularization is incomplete, some of the normal movements of gastrulation occur even in the acellular regions of these mutant embryos.

The pattern of morphogenetic movements at gastrulation is highly asymmetric with respect to both the anterior–posterior and dorsal–ventral axes of the embryo (see Section 4). The primary defect in most mutants that affect the pattern of gastrulation is in the establishment of axial positional information. Maternally expressed genes that are required for the mechanics of morphogenesis have also been identified. For instance, mutants in *concertina* fail to undergo germ band extension, but do not appear to affect the determination of any particular cell types (6).

Future studies should be able to take advantage of maternal-effect mutants to gain insight into the cell biology of the early embryo. Furthermore, it is somewhat arbitrary to consider the early processes of nuclear division, migration and cellularization as separate from the processes of pattern formation, Not only are normal nuclear divisions and cellularization a necessary prerequisite for pattern formation but some pattern generating mechanisms use the same components as those needed for these early processes. For instance, the genes *valois* and *swallow*, which are necessary for generation of the anterior – posterior embryonic pattern (see Section 4), are also needed for normal nuclear migration and cellularization (6,22). As both sets of processes are studied in more detail, the overlap between cell biology and pattern formation in the early embryo will become an important area of research.

4. Establishment of the body axes: the maternal framework

The most exciting recent result of genetic and molecular studies on the role of maternally transcribed genes in directing embryonic development has been the demonstration that gradients of maternal products are used to define position in the early embryo. After decades of arguments about whether such gradients exist, the existence and importance of graded morphogen concentrations have been physically documented in the *Drosophila* embryo.

Embryonic polarity and the gross coordinates along the anterior – posterior and dorsal – ventral axes are defined by the concentration of morphogens along these two axes. These morphogens are encoded by maternally transcribed genes, as are the other molecules that are required to generate the gradients of these morphogens. Local morphogen concentration is measured by zygotic nuclei, which apparently respond by activating or repressing the transcription of specific zygotic genes.

Maternal-effect mutants have been classified as affecting the body axes on the basis of the cuticle phenotype of the differentiated larva, taking advantage of the spatial markers of the larval cuticle (*Figure 3*) (23). Cuticular specializations mark the segmented anterior – posterior pattern, with its progression of head, thoracic and abdominal segments, as well as the different cell types along the dorsal – ventral dimension. Maternal-pattern mutants result in deletions of particular pattern elements, frequently accompanied by the expansion of the remaining structures. For instance, in the 'dorsalized' embryos that result from the absence of the maternal-effect gene *dorsal*, all ventral and lateral pattern elements of the cuticle, including the ventral denticle bands, filzkörper and most head structures, are absent, while the dorsal cuticle is expanded to surround the entire animal (24).

The phenotypes of the mutants that affect the definition of the axes

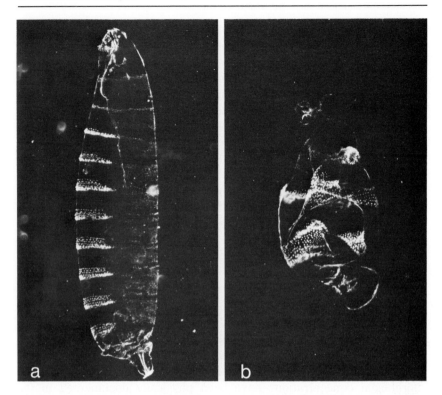

Figure 3. (a) Cuticle of the wild-type first instar larva. The most prominent landmarks of the anterior–posterior pattern are the head skeleton anteriorly (anterior is at the top of the figure) and the thoracic and abdominal ventral denticle bands (ventral is to the left), which lie at the anterior margin of each segment. The ventral denticle bands derive from the lateral and ventrolateral cells of the blastoderm, while dorsal and dorsolateral blastoderm cells give rise to a dorsal cuticle characterized by fine dorsal hairs. **(b)** Cuticle of a differentiated embryo produced by a female of the genotype $T1^{rm9}/T1^{rm9}$. This illustrates the kind of dramatic change in the spatial organization of the larval cuticle that allowed the identification of maternal-effect mutants which change the spatial organization of the embryo. Here the ventral denticle bands encircle the entire animal and the dorsal cuticle is absent. Therefore there is a clear change in the organization of the dorsal–ventral pattern. The phenotype of this embryo has been classified as lateralized, partially on the basis of the expansion of the ventrolaterally derived ventral denticle bands at the expense of dorsal structures (as seen in the cuticular phenotype) and also because the most ventral derivative, the mesoderm, is also absent (as is seen in the pattern of gastrulation and in histological sections).

can be seen not only in the final cuticular pattern of the differentiated embryo, but also very early in development. In most cases, the mutant embryos develop normally until the time of cellularization. At the onset of gastrulation, the first anterior–posterior and dorsal–ventral asymmetries of cell fate in the wild-type embryo can be detected in the pattern of morphogenetic movements (24) and in the local pattern of mitotic divisions (25). In the mutants that affect the embryonic axes, there is a remarkably good correlation between the shifts in cell fate seen in the cuticular pattern and the changes in the pattern of mitoses and cell movements at gastrulation.

The maternal-effect pattern mutants can be cleanly divided into those that affect the anterior – posterior pattern and those that change the dorsal – ventral pattern (24,26,27). The fact that mutants completely alter the pattern in one axis without changing the pattern of the other axis indicates that two independent processes establish the spatial coordinates of the embryo. For both anterior – posterior and dorsal – ventral patterns, axis definition has three phases. The first phase of axis definition occurs during oogenesis, when the architecture of the egg is established. At the time the egg is laid, it appears to have a defined anterior – posterior dimension that is perpendicular to a dorsal – ventral dimension. Although there are anterior – posterior and dorsal – ventral asymmetries in the egg that are normally used to define the polarity of the embryonic axes, embryonic polarity is not fixed with respect to those dimensions. That is, after the egg is laid it is possible to change the polarity of an axis, for instance to cause a head to develop at the posterior end of the animal or to cause ventral structures to develop from the normal dorsal cells, but no experiment has succeeded in altering the orientation of one axis with respect to the other. The second phase of axis definition occurs after egg-laying, when maternally stored components are activated to generate graded distributions of morphogens. For both axes, the second phase occurs early in embryogenesis and precedes the global activation of transcription in zygotic nuclei. In the third phase of axis definition, the local concentration of morphogen is measured and interpreted by zygotic nuclei, resulting in the local activation or repression of specific sets of zyogtic genes at the cellular blastoderm stage.

Although both anterior – posterior and dorsal – ventral axis formation are three-phase processes that involve the creation of morphogen gradients, the formal and molecular natures of the two processes are completely different from each other. The anterior – posterior pattern evolves as the result of the pre-localization of two distinct molecules at the anterior and posterior poles during oogenesis. An anterior gradient develops passively as the result of a localized anterior source and global turnover, while the nature of the posterior organizer is not yet known. The available data suggest that the dorsal – ventral pattern is based on a single graded morphogen, and the source for this morphogen is not pre-localized in the egg. The dorsal – ventral gradient appears to be generated by the interactions among a set of gene products after fertilization.

4.1 The anterior – posterior pattern

The most striking feature of the larval anterior – posterior pattern is the sequence of repeating segmental units. From anterior to posterior there are six head segments, three thoracic segments and eight abdominal segments, and each segment has a unique morphology characteristic of its anterior–posterior position (23). At the extreme ends of the animal lie the unsegmented terminalia, the anterior acron and the posterior telson.

The first maternal-effect mutation that was shown to affect the organization of the spatial pattern of the embryo was *bicaudal* (28,29). Embryos produced by homozygous *bicaudal* mutant females frequently lack all anterior structures and instead form mirror-image symmetrical duplications of the posterior abdomen and telson. This dramatic phenotype provided a graphic demonstration that there is a maternally derived system of information that is necessary for the development of the embryonic pattern. Dominant maternal-effect mutations in two other loci have been described that also produce *bicaudal* embryos, with the same abdominal pattern duplication (30). However, all of the *bicaudal* mutations cause a variable, incompletely penetrant phenotype and none of these mutants represents the absence of gene function. Another maternal-effect mutation, *dicephalic*, causes the production of some double-headed embryos (31) but, like the *bicaudal* mutations, this mutation seems to represent an altered gene activity rather than the loss of an activity necessary to allow production of posterior structures. Thus, although these phenotypes suggest how spatial information might be organized in the egg, there is no compelling genetic evidence that the wild-type product of any of these loci is normally required for the development of the anterior or posterior regions of the embryo.

For maternal-effect mutants in 15 genes the genetic evidence indicates that the function of the wild-type gene product is normally required to define domains of the anterior–posterior pattern (*Table 1*). For each of these genes, multiple recessive mutant alleles give similar phenotypes, as do chromosomal deficiencies. The 15 genes fall into three groups with similar mutant phenotypes, indicating that the generation of the anterior–posterior pattern involves three largely independent pattern-forming pathways (*Figure 4*). The anterior group of genes is required to generate an anterior organizing center that defines an anterior domain in the embryo, roughly head plus thorax, and the posterior group of genes is necessary to promote the development of the abdomen. The genes of the terminal group draw the boundaries between the unsegmented terminalia and the segmented body of the animal. It is these simple subdivisions that are used by the embryonic nuclei to generate the detailed anterior–posterior pattern. The description of the nature of these three pathways and the function of these genes in the next several sections is based largely on the proposals of Nüsslein-Volhard, Frohnhöfer and Lehmann (26).

4.1.1 The anterior organizing center

Three maternally expressed genes that are necessary for the production of anterior structures of the larva have been identified. In the absence of *bicoid, exuperantia* or *swallow* gene activity in the mother, the embryos lack anterior structures (22,32,33). In *exuperantia* and *swallow* mutant embryos, anterior structures of the head are deleted, while *bicoid* embryos show a more extreme phenotype, lacking both head and thorax.

In the absence of the activity of one of the anterior group genes, as anterior structures are lost, the region of the presumptive abdomen is increased in size, expanding anteriorly (*Figure 5*). This shift in the fate map can be seen in the pattern of morphogenetic movements at gastrulation (32). In the wild-type embryo the head fold occurs at 65% egg length (0% = posterior pole; 100% = anterior pole) at the onset of gastrulation. Weak *bicoid* alleles cause the head fold to be shifted anteriorly and in strong alleles it is absent. The fate map shifts can also be visualized in the pattern of expression of the zygotically expressed segmentation gene *fushi tarazu* (*ftz*) (22). In the wild-type embryo, *ftz* is expressed at the cellular blastoderm stage in seven stripes spaced along the anterior – posterior axis, with the most anterior stripe ending at 65% egg length (see Chapter 2). In *swallow, exuperantia* and *bicoid* mutant embryos, the *ftz* stripes are shifted anteriorly, with a more extreme shift seen in *bicoid* embryos than in *swallow* or *exuperantia* embryos. Thus the activity of the anterior group genes defines the domain of the embryo that will give rise to head and thorax, and thereby defines the anterior boundary of the abdomen.

With sets of genes that result in similar phenotypes, it can be difficult to assign specific functions to specific genes. A powerful tool for assigning functions to maternally active genes has been the ability to rescue mutant phenotypes by the injection of wild-type cytoplasm. In such experiments, cytoplasm from a particular position of a young wild-type embryo is removed with a microinjection needle and is then injected at a defined position into a young, pre-cellular embryo produced by a female carrying a maternal-effect mutation. For many maternal-effect mutants, the wild-type cytoplasm partially or completely restores the normal pattern of development. Rescue can be scored in the pattern of the differentiated cuticle at the end of embryogenesis, or, in some cases, in the pattern of morphogenetic movements at gastrulation.

In the case of anterior group genes, cytoplasmic rescue experiments established that it is the *bicoid* gene product that defines the anterior signal (32). The *bicoid* mutant phenotype can be rescued by injecting embryos from *bicoid* mothers with cytoplasm from wild-type embryos. To rescue the mutant phenotype, it is necessary to transplant wild-type cytoplasm from the anterior tip of the wild-type donor into the anterior tip of the *bicoid* host, indicating that the *bicoid* rescuing activity is both localized and required at the anterior pole of the embryo.

In addition to its ability to restore a normal pattern to *bicoid* mutant embryos, the *bicoid*[+] activity has the ability to induce head structures

Table 1. Genes with maternal-effect alleles that alter the spatial pattern of embryonic development

Gene	Phenotype	Proposed function and structure	References
Anterior group			
bicoid (bcd)	Head and thorax deleted, replaced by inverted telson	Graded anterior determinant; homeodomain-containing	32,35–37
exuperantia (exu)	Anterior head structures deleted	Anchors bicoid RNA	22,33
swallow (swa)	Anterior head structures deleted	Anchors bicoid RNA	22,43
Posterior group			
tudor (tud)	No abdomen, no pole cells		33,80
oskar (osk)	No abdomen, no pole cells		44
vasa (vas)	No abdomen, no pole cells; oogenesis defective		33,47
valois (val)	No abdomen, no pole cells; cellularization defect		33
staufen (stau)	No abdomen, no pole cells; head defect		33
nanos (nos)	No abdomen	Posterior signal	26
pumilio (pum)	No abdomen	Transport posterior signal to abdomen	48
Terminal group			
torso (tor)	No terminalia		33,52
trunk (trk)	No terminalia		33
torsolike (tsl)	No terminalia		26,49
fs(1)Nasrat[fs(1)N]	No terminalia; collapsed eggs		51
fs(1)pole hole[fs(1)ph]	No terminalia; collapsed eggs		50
Other anterior – posterior pattern genes			
caudal[a] (cad)	Reduced abdomen with abnormal segmentation	Homeodomain-containing	42
bicaudal (bic)	Double abdomen		28,29
Bicaudal[C] (Bic C)	Double abdomen		30

Bicaudal[D] *(Bic D)*	Double abdomen		30
dicephalic (dic)	Double heads; some double abdomens		31
Egg shape and embryonic pattern			
fs(1)K10	Dorsalized egg and embryo	Nuclear protein	54,59
gurken (grk)	Ventralized egg and embryo		55
torpedo (top)	Ventralized egg and embryo		55
Dorsal group			
dorsal (dl)	Dorsalized	Graded morphogen; nuclear protein	24,64,65
windbeutel (wbl)	Dorsalized		6
gastrulation defective (gd)	Dorsalized		5,81
nudel (ndl)	Dorsalized		60
pipe (pip)	Dorsalized		60
tube (tub)	Dorsalized		60
snake (snk)	Dorsalized	Protease	71
easter (ea)	Dorsalized	Protease	b
Toll (T1)	Dorsalized	Organizes polarity; transmembrane protein	62,66,72
spätzle (spz)	Dorsalized		60
pelle (pll)	Dorsalized		60,82
Other dorsal–ventral pattern genes			
cactus (cac)	Ventralized		6
Neurogenic			
almondex (amx)[a]	Neuralized		75
pecanex[a]	Neuralized		50
mat (2) Notchlike[a]	Neuralized		6

[a]Male-rescuable maternal-effect mutants. Extreme phenotype only when both mother and zygote are mutant.
[b]R.Chasan and K.V.Anderson, submitted.

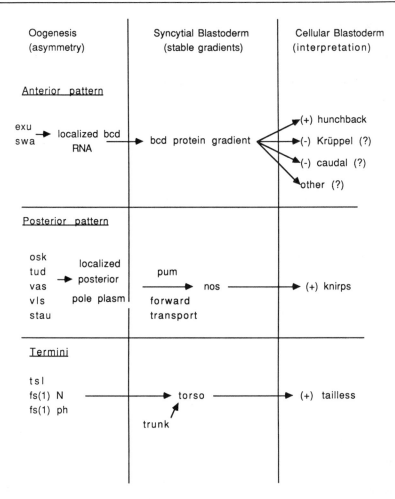

Figure 4.   The three independent genetic pathways that organize the anterior – posterior pattern of the embryo. In each of three pathways, the initial asymmetry is set up during oogenesis, the pattern is organized after fertilization by maternally encoded products, and the pattern is realized by the localized expression of genes in the zygote. The arrows are meant to indicate the genetic hierarchy and not a biochemical pathway.

at ectopic sites in the embryo (32). For instance, if cytoplasm from the anterior tip of the wild-type embryo is injected into the middle, 50% egg length, of a *bicoid* mutant embryo, head structures develop at that site, and thoracic structures form mirror-symmetrically away from the injection site. Thus the *bicoid* gene product is not only necessary for the production of anterior structures, it is sufficient to promote their development.

Cytoplasm from the anterior tip of *swallow* or *exuperantia* embryos rescues *bicoid* embryos only weakly (22), indicating that the *bicoid*-rescuing activity in the wild-type embryo depends on *exuperantia* and

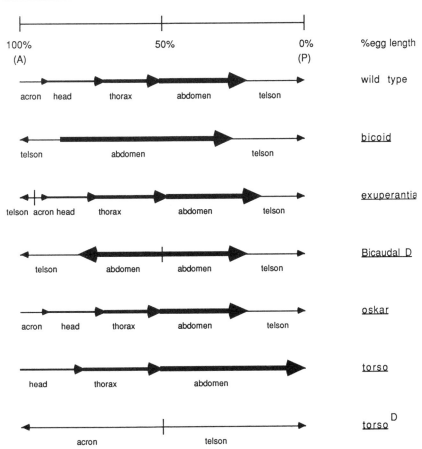

Figure 5. Changes in the blastoderm fate map in maternal-effect mutants with altered anterior – posterior patterns. In the wild-type embryo, blastoderm cells choose to develop as acron, head, thorax, abdomen, or telson as a function of their anterior – posterior position in the embryo. In some mutants that fate map is shifted. For instance, in *bicoid* mutant embryos, cells that would normally give rise to head and thorax contribute to the abdomen. In other mutants that affect the development of a particular anterior – posterior region, the fate map is not shifted. In *oskar* mutant embryos, for instance, although the abdomen fails to differentiate, the cells of the normal abdominal region do not contribute to thorax or telson. The fate maps drawn here are based on the pattern of morphogenetic movements at gastrulation and on the pattern of expression of the segmentation gene *ftz* at the blastoderm stage. Since *ftz* is not expressed in the head, the fate maps drawn in that region are probably somewhat inaccurate. Arrowheads indicate the polarity of the pattern within a region. A, anterior pole; P, posterior pole.

swallow. This result suggested that the defects seen in *exuperantia* and *swallow* embryos are the result of lowered *bicoid* activity. Thus the injection experiments showed that *bicoid* is necessary for anterior development, is sufficient to promote anterior development at ectopic sites and, among the known genes, is most directly responsible for promoting head development.

DNA clones including the *bicoid* gene were isolated because *bicoid* lies within the *Antennapedia* complex, a cluster of homeobox-containing genes that had previously been cloned (34,35; see also Chapter 2). A region encoding a maternally expressed transcript was positively identified as the *bicoid* gene by the ability of that DNA to complement the *bicoid* mutant phenotype after reintroduction into the genome by P-element-mediated transformation (35). Using cloned *bicoid* DNA probes and antibodies directed against a *bicoid* fusion protein synthesized in *Eschericha coli*, it has been confirmed that it is the product of the *bicoid* gene that promotes anterior development (36,37). In the wild-type egg, the *bicoid* transcript is localized at the anterior pole at the time of fertilization. At the time the egg is laid, no *bicoid* protein has been synthesized. Egg-laying (not fertilization) activates translation of the *bicoid* RNA. Two hours after egg-laying, at the syncytial blastoderm stage, a stable gradient of *bicoid* protein has been generated, extending from a high point at the anterior pole to undetectable levels at 30% egg length. The shape of the gradient can be explained if it is assumed that there is a localized source of protein (the anteriorly localized *bicoid* RNA) and uniform protein turnover throughout the embryo, with a half-life of the *bicoid* protein of less than 30 min (36).

The *bicoid* protein is found in embryonic nuclei and includes a homeobox domain (36,37), so it is likely to act by directly regulating the transcription of genes in zygotic nuclei in a concentration-dependent manner. Above certain concentration thresholds, *bicoid* protein presumably binds upstream of target genes and either positively or negatively regulates their transcription. The patterns of expression of a number of cloned zygotic genes required for segmentation have been compared in wild-type and *bicoid* embryos (38–40). The absence of *bicoid* causes a change in the spatial pattern of expression of all zygotic segmentation genes examined. While these results illustrate the profound effects of *bicoid* on the anterior–posterior pattern, these data alone do not distinguish between direct and indirect consequences of *bicoid* protein action.

Many features of the *bicoid* phenotype can be explained if it is hypothesized that above a certain threshold concentration, corresponding to the *bicoid* concentration found at 65% egg length, the *bicoid* protein turns on the transcription of the *hunchback* gene. *Hunchback* is a zygotically expressed segmentation gene required for the development of most of the anterior segments that require *bicoid* (2). *Hunchback* is normally transcribed beginning early in the syncytial blastoderm stage in the anterior third of the embryo, and is apparently not transcribed in *bicoid* mutant embryos (40). Since the *hunchback* gene has also been cloned, the hypothesis that the *bicoid* protein binds to regulatory regions of the *hunchback* gene to activate transcription can now be tested directly.

Since the region affected in *bicoid* mutants extends both anterior to and posterior to the region deleted in embryos that lack zygotic expression of *hunchback*, there must be other gene targets for *bicoid* regulation. That is, there must be a series of threshold concentrations of *bicoid* protein that can be detected by zygotic nuclei, and above each threshold specific zygotic genes are activated or repressed. It has been suggested that *bicoid* might activate the zygotic genes *tailless* or *giant* at the anterior end of the embryo and that *bicoid* could repress the zygotic gene *Krüppel* in the thoracic region (26). It has also been proposed that *bicoid* represses the production of the homeodomain-containing protein *caudal* (41). *caudal* was first identified molecularly as a maternally expressed homeobox-containing gene. Subsequent genetic analysis showed that in the absence of both maternal and zygotic expression of *caudal*, abdominal segmentation is disrupted (42). The *caudal* protein forms a posterior-to-anterior gradient at the syncytial blastoderm stage. Since the *caudal* RNA is uniformly distributed at egg-laying and the *caudal* protein gradient can form in the absence of zygotic transcription, the gradient must be formed by a post-transcriptional mechanism. *caudal* protein is present at high concentration throughout *bicoid* embryos, but the gradient is normal in mutants for the early possible targets of *bicoid*, *hunchback* and *Krüppel* (41). Thus it is reasonable to propose that *bicoid* acts directly to inhibit the production of the *caudal* protein. This would have to occur at a post-transcriptional level, suggesting a novel mechanism of action for the homeodomain-containing protein *bicoid*.

Rescue experiments had suggested that *bicoid* activity depends on the activities of the *swallow* and *exuperantia* genes. Using cloned *bicoid* probes, it has been found that in the absence of either of those two gene products, the *bicoid* RNA fails to be localized at the anterior end of the embryo. In oocytes of *swallow* or *exuperantia* females, *bicoid* RNA is present, but is distributed evenly (35). After fertilization, a shallow RNA gradient is seen in these mutants, with a decreased *bicoid* RNA concentration at the posterior pole. This decrease in the stability of *bicoid* RNA posteriorly is mediated by the posterior organizing center (see below), since it is not seen in *exuperantia – vasa* or *exuperantia – oskar* double mutants (35). Since no *bicoid* RNA is present in the posterior part of the wild-type embryo, the biological significance of this activity of the posterior organizing center is not clear.

The *swallow-* and *exuperantia*-dependent mechanism which leads to the localization of *bicoid* RNA has not yet been elucidated, but a simple model has been proposed (35). The anterior tip of the oocyte is the end of the oocyte closest to the nurse cells. *Bicoid* RNA is synthesized in the nurse cells and, as soon as it is transported into the oocyte, becomes localized at the anterior pole. It is assumed that the *bicoid* RNA becomes associated with a specific binding protein in the nurse cells and that the *bicoid* RNA-binding protein complex can bind to a component of the oocyte cortical cytoplasm. The *bicoid* RNA-binding protein complex would then bind to the first site that it is exposed to which exhibits cortical specialization, the anterior tip of the oocyte. This model supposes the existence of a specific *bicoid* RNA-binding protein and of a specialization of the oocyte cortex that is absent from the nurse cells. If *swallow* and *exuperantia* encode components of this process, it seems likely that *swallow* might be necessary for the egg's cortical specializations, since mutations in *swallow* are also associated with variable defects in cellularization of the blastoderm (43). Since the only effect of mutations in *exuperantia* is to block the localization of the *bicoid* RNA, it is reasonable to propose that the function of the wild-type product of *exuperantia* is to bind the *bicoid* RNA and then mediate its association with the oocyte cytoskeleton. Thus a relatively simple mechanism could localize the *bicoid* RNA and this localization has profound consequences for embryonic development.

4.1.2 The posterior organizing center

In the absence of the anterior organizing center in embryos from *bicoid* mutant mothers, the organization and polarity of the abdomen is nearly normal (32), indicating that some other system organizes the posterior half of the animal. Mutations in seven different maternally active genes, *oskar, tudor, vasa, valois, staufen, nanos,* and *pumilio*, result in the loss of abdomen and thus define a pathway necessary to organize the posterior embryonic pattern (*Figure 4*) (26,33,44).

The posterior organizing activity differs fundamentally from the anterior organizing center in that it does not define the boundaries of a region of the embryonic fate map (*Figure 5*). Mutations in the posterior group genes do not cause a change in the fate map: the cells of the presumptive abdominal region do not differentiate correctly, but there is no expansion of the anlage for head, thorax or telson (44). The normal fate map in these mutants has been documented both by analyzing the position of cell shape changes at gastrulation (44) and by defining the region in which the zygotic gene *ftz* is expressed in the mutant embryos (38). Thus the domain in the wild-type embryo where the posterior organizing center's activity is required is not defined by the posterior activity itself. Instead, the anterior boundary of the abdomen appears to be defined by *bicoid*, and its posterior boundary by *torso* (see Section 4.1.3). Thus despite the fact that the posterior activity has the capacity to inhibit *bicoid* in experimental situations (32), apparently by destabilizing the *bicoid* RNA (35), there is no need to invoke an interaction between the two polar organizers in normal development.

Cytoplasmic rescue experiments have shown that the posterior organizing center is localized at the posterior pole and not in the presumptive abdominal region as one might have guessed from the abdominal defect (44). The abdominal segmentation defect in mutants in any of the seven loci can be rescued by the injection of wild-type cytoplasm (26). Rescue is obtained only when wild-type cytoplasm is injected into the presumptive abdominal region of the mutant embryo (44). In addition, rescue is only seen when the wild-type cytoplasm is taken from the posterior pole of the donor and not from the presumptive abdominal region. Thus a substance (or substances) sequestered at the posterior pole is required at a more anterior site to allow abdominal development.

The posterior pole cytoplasm that contains the abdomen-rescuing activity is the only morphologically specialized cytoplasm visible in the egg when it is laid (8). This cytoplasm is relatively yolk-free, enriched in mitochondria, and contains a kind of specialized ribonucleoprotein complex, the polar granules. The pole plasm is incorporated into the first cells made in the embryo, the pole cells, which are the precursors to the germ line cells. Transplantation experiments have shown that the specialized pole plasm can induce the formation of germ cells at other sites in the embryo, indicating that the pole plasm contains some sort of germ line determinant (45). Thus two activities are sequestered in the pole plasm, a germ line determinant and a component necessary for the organization of the posterior half of the body plan.

The production and localization of the germ line determinant and the abdominal organizer are mutually dependent processes. Mutations in five of the posterior group genes, *oskar, tudor, vasa, valois* and *staufen*, cause a dual phenotype (26,33,44): embryos from females mutant for any one of these genes lack an abdomen, have no specialized posterior pole plasm and do not make pole cells. A simple explanation for this dual phenotype is that these gene products are required neither for the production of a signal for abdominal development nor for germ line development, but are instead necessary to localize both signals at the posterior pole. Rescue experiments in which posterior pole cytoplasm from embryos mutant for one locus is injected into the abdominal anlage of embryos mutant for another posterior group gene are consistent with this interpretation. In such cytoplasmic complementation experiments, posterior cytoplasm from any of the five mutants fails to rescue the mutants in other loci (46). Thus the members of this group define steps in a dependent pathway, in which the synthesis or stability of some components depends on the activity of the other gene products.

Each of these five genes is required to allow the localization of the two signals at the posterior pole. In contrast to the localization of *bicoid* RNA, this posterior localization process cannot be completely passive, since the contents of the nurse cells enter the oocyte from the anterior tip. Perhaps these five genes encode a structure that specifically transports and binds molecules to the posterior pole. Alternatively, the substances that become localized to the posterior pole might have some inherent affinity for both poles of the oocyte, and cannot bind at the anterior pole because a specific inhibitor that is loaded into the egg before the posterior substances competitively blocks binding to the anterior pole. There is some underlying similarity between the two poles since rescue experiments have shown that in *Bicaudal*[D] embryos the posterior organizing activity is localized at both the posterior and anterior poles of the embryo (44).

The mutant phenotypes associated with these five genes are not exactly identical, suggesting that they serve different functions (33). Mutations in *staufen* cause a head defect in addition to the abdominal defect. Mutations in *valois* frequently cause cellularization defects. Null alleles of *vasa* cause the arrest of oogenesis at an early stage, before the pole plasm is visible (47). The *vasa* gene has been cloned and the sequence of its predicted protein product is similar, across almost its entire length, to murine translation initiation factor, eIF-4A. The *vasa* sequence includes a consensus ATP-binding site and other features associated with nucleic acid helicases (47). Molecular studies on the other genes should make it possible to define the role of this putative nucleic acid-binding protein and reveal how the posterior group genes act together to create the posterior organizing center. The two remaining genes of the posterior group, *nanos* and *pumilio*, affect only abdominal segmentation and not germ line determination. Embryos from *nanos* or *pumilio* mutant females lack an abdomen, but the posterior pole plasm in these embryos appears normal, pole cells form and the pole cells can function as germ line precursors (26,48). Thus these two gene products are not required for the assembly or localization of the posterior pole plasm, but are specifically required for the production of the abdomen.

Rescue experiments have suggested that the *pumilio* gene product is specifically required to transport a signal from the posterior pole plasm to the presumptive abdominal region (48). The posterior pole plasm of *pumilio* embryos does contain the signal required for abdominal development, since if the pole plasm from a *pumilio* mutant embryo is transplanted into the presumptive abdominal region of another *pumilio* embryo, the development of the abdomen is rescued. The hypothesis that *pumilio* is required for anteriorward transport is also supported by double mutant studies (48). In embryos from females that are mutant for both *pumilio* and *torso*, which causes a deletion of the telson (the region which lies between the pole plasm and the presumptive abdomen), abdominal segmentation is restored. Thus by shortening the distance between the posterior source and the abdominal target, the requirement for *pumilio* is by-passed.

The product of the *nanos* gene is the only candidate for the signal that actually promotes development of the abdomen. Embryos from *nanos* mothers have normal pole cells, and lack abdomen-promoting activity. If all members of the posterior group of genes have been identified, then *nanos* is the only maternal gene required for the complex task of organizing the entire abdominal region. The *nanos* product probably acts by influencing the zygotic expression of *knirps*, a zygotically expressed gene which is also required for abdominal development (2). The *nanos*-dependent posterior signal that organizes the abdomen must differ fundamentally from the *bicoid* signal that directs anterior development. Two lines of evidence indicate that *nanos* cannot promote a simple posterior-to-anterior gradient. First, the region that requires the highest level of the posterior organizing activity is not the most posterior abdominal pattern element, but instead the center of the abdomen (44). Second, when posterior pole plasm is injected into the central region of embryos that lack both anterior and posterior organizers (e.g. *bicoid – oskar* double mutants), bicaudal-type embryos form with both poles of the embryo forming posterior ends (26). That is, although in the wild-type animal the most posterior structures form nearest the source of *nanos* activity, in the injected animals the most posterior structures develop at the greatest distance from the source of *nanos* activity.

There is no good candidate for a gene that encodes the germ line determinant. Mutations in such a gene would be expected to allow normal abdominal development and could be identified only on the basis of a grandchildless phenotype. Although screens for grandchildless mutants have been performed, these screens require four generations and because of their laborious nature have been carried out only on a limited scale. Future studies may identify the germ line determinant either on the basis of a grandchildless mutant phenotype or on the basis of a biochemical interaction with the product of one of the posterior group genes.

4.1.3 Terminal genes

In the absence of both the anterior and posterior organizing centers, embryos still develop some anterior – posterior pattern (26). Embryos from *bicoid – oskar* double mutant females, for instance, produce two telsons with mirror-image symmetry, one at the posterior pole and the other at the anterior pole. The activity of a third set of genes, the terminal group, must also be abolished to remove all anterior – posterior embryonic pattern. Mutations in the terminal genes, *torso, trunk, torsolike*, fs(1) *Nasrat* and fs(1) *pole hole*, result in the loss of structures of the non-segmented extremities of the animal, the acron and the telson (33,34,49 – 51). Analyses of blastoderm fate maps in these mutant embryos have shown that the segmented region of the fate map is expanded to the tips of the embryo (*Figure 5*) (51,52). Thus the function of the wild-type products of these genes is to allow the production of terminalia and to define the limits of the segmented domain of the body.

The underlying similarity between anterior and posterior terminalia suggested by these phenotypes is even more apparent when considering the *bicoid* phenotype. Embryos from *bicoid* mutant mothers lack head and thorax, and at the anterior end the most terminal region develops as an extra telson with reverse polarity (32). Thus in the absence of *bicoid* activity, the two ends of the animal develop identically. The development of the acron, then, must be the result of the combinatorial action of *bicoid* and the terminal genes.

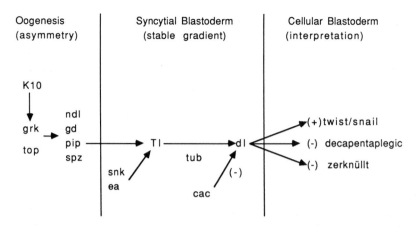

Figure 6.   The genetic pathway that organizes the dorsal – ventral pattern of the embryo. The pathway shown here, based on double mutant phenotypes, illustrates some features of the genetic hierarchy among the genes required for dorsal – ventral pattern organization. Arrows are not meant to imply a biochemical pathway. Some proposed features of the pathway, including a positive autocatalytic loop including *Toll* and perhaps *easter* (62; R.Chasan and K.V.Anderson, submitted), have not been included.

The *torso* gene appears to play a key role among the terminal genes (*Figure 4*) (52). Dominant gain-of-function alleles of *torso* have been isolated which, instead of deleting the terminal regions, cause the termini to be expanded at the expense of the segmented portion of the body. The existence of such dominant gain-of-function alleles with phenotypes opposite to loss-of-function alleles is a hallmark of genes that control developmental decisions. Not only is the activity of such a gene required, but excessive or inappropriate activity of the gene is sufficient to drive cells along an inappropriate pathway. The opposing phenotype of the loss-of-function and gain-of-function alleles suggests that *torso* activity can drive cells to differentiate as terminalia, and in the dominant alleles, the mechanisms that normally restrict *torso* activity to the poles are inoperative. The *torso* phenotype can be rescued by the injection of wild-type cytoplasm into the terminal regions of the embryo, demonstrating a local requirement for the *torso* gene product (*Figure 6*) (52). However, the *torso*-rescuing activity is uniformly distributed in the wild-type embryo, being present both at the ends and in the middle of the embryo. The local requirement and global presence of *torso*-rescuing activity, together with the phenotypes of the dominant alleles, imply that in normal development a homogeneously distributed precursor form of *torso* is activated only at the poles. The other terminal genes may be required to allow normal *torso* activity. While the temperature-sensitive periods of temperature-sensitive alleles of both *torso* and *trunk* are exclusively after fertilization (33), at the syncytial blastoderm stage, the other genes of this class probably act before fertilization. The temperature-sensitive period of a *torsolike* allele takes place during oogenesis (49). Null alleles of fs(1) *Nasrat* and fs(1) *pole hole* both cause the production of collapsed eggs (50,51), perhaps a more severe form of the same molecular defect that causes the absence of terminalia in the embryo. Thus if the production of the terminalia is a simple pathway, *torsolike*, fs(1) *Nasrat*, and fs(1) *pole hole* act upstream of *torso*.

The zygotically active gene *tailless* has a phenotype similar to the maternal terminal genes, with structures from both extremities deleted (53). Double mutants of dominant maternal *torso* alleles with zygotic *tailless* alleles look like the embryos caused by *tailless* single mutants: they make the segmented region of the body. This suggests that ectopically active *torso* cannot promote the formation of terminalia in the absence of the *tailless* gene product, and that implies that the wild-type *torso* product acts by activating the zygotic *tailless* gene in the terminal regions of the embryo.

4.2 The dorsal – ventral pattern

At the same time blastoderm cells measure their anterior – posterior position, they also assess their dorsal – ventral position and differentiate into different cell types as a function of that position (23). The ventral-most 20% of the cells invaginate in the ventral furrow at gastrulation to form the mesoderm. The lateral and ventrolateral cells compose the neurogenic ectoderm, which gives rise to the ventral nerve cord and the ventral epidermis. Dorsal and dorsolateral cells give rise to the dorsal epidermis, while the extreme dorsal cells make the extraembryonic amnioserosa.

The maternal-effect mutations that disrupt the embryonic dorsal – ventral pattern affect both the final differentiated fate of cells and also the early morphogenetic movements of gastrulation (24). Gastrulation in wild-type embryos is highly asymmetric dorsoventrally. The first event of gastrulation is the invagination of the cells on the ventral midline in the ventral furrow to generate the presumptive mesoderm. The most prominent subsequent morphogenetic movement is germ band extension, in which the anterior – posterior axis of the future embryo elongates along the dorsal side of the embryo. The effect of mutants on the differentiated pattern can be accurately predicted by the pattern of early morphogenetic movements. For instance, the strongly dorsalized embryos produced by females that lack activity of the *dorsal* gene show no dorsal – ventral asymmetry at gastrulation, do not make a ventral furrow and do not extend the germ band. When such embryos differentiate, they lack mesoderm, nerve cord and ventral epidermis. Thus, as with the anterior – posterior pattern, the basic features of the dorsal – ventral pattern have been established and interpreted by this early point in development under the direction of maternally stored products.

Two classes of maternally active genes are required to generate the embryonic dorsal–ventral pattern (*Table 1*) (*Figure 6*) (27). Three genes act during oogenesis to organize the spatial pattern of both the eggshell and the embryo that develops within that eggshell. After the stage when eggshell and embryonic polarity are coupled, the products of an additional 12 genes are required to organize the dorsal–ventral pattern of the embryo itself. The available data suggest that the gene products necessary for the embryonic dorsal–ventral pattern are not pre-localized within the egg and that dorsal–ventral asymmetry arises as a consequence of specific protein–protein interactions in the blastoderm embryo. The consequence of these epigenetic interactions is the generation of a gradient of the protein product of the *dorsal* gene.

4.2.1 Dorsal–ventral asymmetry during oogenesis

The wild-type eggshell is flat on one side, which corresponds to the presumptive dorsal side of the embryo, and curved on the presumptive ventral side (8). The paired chorionic appendages, specializations of the eggshell that promote gas exchange, are located on the anterior dorsal surface of the eggshell. Egg shape and appendage position reflect the arrangement and behavior of the follicle cells that surround the oocyte. For instance, the dorsal chorionic appendages are synthesized by a particular group of follicle cells that migrate while secreting chorion proteins in order to produce the elongated final structure.

At the time the egg is laid, there must be some asymmetric cue that normally reliably causes the ventral side of the embryonic pattern to coincide with the curved, ventral side of the eggshell. A set of genes that may define this asymmetric cue has been identified. Mutations in these genes affect both the shape and polarity of the egg and the pattern of development of the embryo contained in the egg. Thus embryonic pattern must be initially coupled to eggshell pattern, either because the eggshell contains components that are used as chemical cues to orientate the embryonic pattern or because the shape of the egg itself provides a physical asymmetry used to organize embryonic pattern.

Mutations in three genes that change the normal asymmetry of the eggshell and also cause corresponding shifts in the embryonic dorsal–ventral pattern, fs(1)*K10*, *gurken* and *torpedo*, have been characterized in some detail (54,55). Females homozygous for mutations in fs(1)*K10* produce dorsalized eggshells: the egg becomes barrel-shaped, flattened both dorsally and ventrally, and the chorionic appendages are present as a ring around the entire dorsal–ventral circumference. In contrast, *gurken* and *torpedo* mutations ventralize the eggshell and egg shape, producing long, spindle-shaped eggs that are curved on all sides and that lack chorionic appendages. Although the egg shape changes found in these mutants reduce the likelihood that the eggs will be successfully fertilized through the micropyle (also an eggshell specialization), when the mutant eggs are fertilized, the embryonic pattern in *K10* embryos is partially dorsalized and that of *torpedo* and *gurken* embryos is ventralized.

Studies of mutants in these genes illustrate the importance of communication among the three ovarian cell types in the production of the egg. These mutations affect cells of two very different lineages: the oocyte and the nurse cells are germ line derivatives, and the follicle cells, which build the eggshell, are of mesodermal origin. Genetic mosaics in which the germ line cells, nurse cells and oocyte, were mutant while the somatic cells, including the follicle cells, were wild-type have been constructed for mutants in each of these three loci. In the case of *K10* and *gurken*, such mosaic females produce the same mutant eggs as when all cells in the female are mutant, showing that the phenotype is determined only by the genotype of the nurse cells and oocyte (55,56). Therefore, in the wild-type ovary, the germ line cells must produce some *K10*- and *gurken*-dependent signal which directs the behavior of the neighboring follicle cells. Similar mosaic females carrying *torpedo* mutations in the germ line but not in the soma produce wild-type eggs and embryos, while the reciprocal genetic mosaics in which the germ line is wild-type but the somatic cells are mutant show the typical *torpedo* ventralized phenotype (55). Thus the *torpedo* product made in somatic cells, presumably the follicle cells, affects the subsequent development of the neighboring oocyte, in addition to the behavior of the follicle cells themselves. The phenotypes of double mutants have been used to order partially the activity of these genes (55). Females that carry mutations in both *K10* and *gurken* or both *K10* and *torpedo* produce the ventralized eggs and embryos characteristic of *gurken* or *torpedo* alone, suggesting that *torpedo* and *gurken* act downstream of *K10*. Double mutants of *torpedo* or *gurken* with *dorsal* cause the production of ventralized eggshells in which dorsalized embryos develop. Therefore the effects of *torpedo* and *gurken* on the dorsal–ventral pattern of the embryo require the wild-type activity of *dorsal*.

The *K10* locus has been cloned by microdissection and microcloning of the band from salivary gland polytene chromosomes known to include the *K10* locus, and its identity has been confirmed by transformation (57). As predicted by the genetic mosaics, the *K10* RNA and protein are not present in the follicle cells. The surprising result of the molecular analysis was that although the vast majority of the RNA and protein found in the oocyte is synthesized in the nurse cells, the *K10* transcript is synthesized by the oocyte nucleus, and the *K10* protein accumulates in the oocyte nucleus (58,59). The predicted *K10* protein sequence includes a region that resembles a helix-turn-helix motif characteristic of some DNA-binding proteins (although it is clearly not a homeobox) (59), suggesting that the *K10* protein could regulate the transcription of other genes in the oocyte nucleus.

On the basis of this set of mutants, a possible chain of events that defines eggshell and embryonic polarity can be proposed. The *K10* gene product in the oocyte nucleus regulates the synthesis of other oocyte-specific transcripts. Since the oocyte nucleus lies dorsally in the oocyte, these transcripts might then influence the behavior of the nearest, dorsal, follicle cells. This interaction between the oocyte and follicle cells might require the *gurken* product in the oocyte and the *torpedo* product in the follicle cells. The spatially asymmetric activity of the follicle cells may then cause a chemical or physical asymmetry that is interpreted by the embryo as a spatial trigger which then normally defines the dorsal – ventral axis of the embryo.

4.2.2 Generation of the embryonic dorsal – ventral pattern

In addition to the asymmetric cue provided by the eggshell, the products of 12 maternally active genes are required to organize the dorsal – ventral pattern of the embryo itself. In the absence of any one of 11 dorsal group genes in the mother (*dorsal, Toll, gastrulation defective, windbeutel, nudel, pipe, tube, snake, easter, spätzle* and *pelle*), egg shape is normal but cells at all positions in the embryos develop like the dorsal cells of the wild-type embryo (60). Mutants in one gene, *cactus*, have an opposing phenotype. The absence of *cactus* activity in the mother leads to the production of 'ventralized' embryos in which dorsal structures are absent from the embryo and there is a corresponding expansion of ventrolateral structures (6).

The phenotypes of mutants in the 11 dorsal group genes indicate that the entire dorsal – ventral pattern is based on a graded continuum of information (24,60). Absence of the activity of any one of these genes produces an embryo in which all cells generate dorsal structures. Partial loss-of-function alleles, which have been isolated for in most loci (60; K.V.Anderson and S.Wasserman, unpublished results) allow the production of lateral as well as dorsal structures. Thus the highest concentration of these gene products is required to allow the production of ventral structures and intermediate concentrations for lateral structures. Fate-mapping studies have shown that in the partial loss-of-function mutants the normal fate map is shifted (*Figure 7*) (61). For instance, ablation of midventral cells in the wild-type embryo, which give rise to mesoderm, does not cause visible defects in the larval cuticle. However, ablation of midventral cells in a partially dorsalized embryo causes defects in the ventral denticle bands of the cuticle. Thus the cells at a midventral position in the mutant behave like the ventrolateral cells of the wild-type embryo. The simplest interpretation of these allelic series and fate-mapping studies is that these genes allow the production of a morphogen that is normally distributed in a gradient, with its highest concentration ventrally. Blastoderm cells read the local morphogen concentration and differentiate accordingly. When one of these gene products becomes limiting in a partial loss-of-function mutant, the total amount of the graded morphogen is reduced, so when cells read the local morphogen

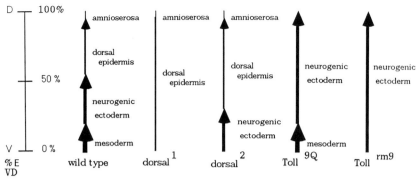

Figure 7. Changes in the blastoderm fate map in maternal-effect mutants with altered dorsal – ventral patterns. In the wild-type embryo, cells choose developmental fates as a function of their dorsal – ventral position in the embryo. Based on the pattern of morphogenetic movements at gastrulation and the pattern of expression of the gene *zerknült* (67), the dorsal – ventral maternal-effect mutants all cause shifts in the blastoderm fate map. *dorsal*[1] is a strong, apparent null allele, *dorsal*[2] is a weak, partially dorsalizing allele, *Toll*[9Q] is a dominant, ventralizing allele, *Toll*[rm9] is an unusual recessive lateralizing allele. Arrowheads indicate polarity within a region. E, egg circumference; V, ventral; D, dorsal.

concentration they respond like the cells that lie at more dorsal positions in the wild-type embryo (24).

Based on the analysis of temperature-sensitive periods and on the sensitivity of mutant embryos to rescue by the injection of wild-type cytoplasm after fertilization, embryonic dorsal – ventral pattern formation appears to involve two sequential steps. Five genes, *gastrulation defective, nudel, pipe, spätzle* and *windbeutel*, appear to act primarily during the late stages of oogenesis, since they are associated with temperature-sensitive periods beginning late in oogenesis and ending shortly after fertilization, and mutants are either not rescuable, or only weakly rescuable, by the injection of wild-type cytoplasm. By the same criteria, *tube, snake, easter, pelle, Toll* and *dorsal*, as well as *cactus*, appear to act exclusively after fertilization during the syncytial blastoderm stage.

Simple double mutants among loss-of-function alleles of the dorsal group genes have not been informative about the order of action of the gene products, since both single and double mutant phenotypes are dorsalized. Double mutants of recessive dorsalizing alleles with ventralizing mutations have been more helpful. Ten dominant, gain-of-function alleles of *Toll* that ventralize the embryonic pattern have been isolated (62; M.Erdélyi and J.Szabad, submitted; K.V.Anderson, S.Wasserman and P.Hecht, unpublished results), suggesting that the activity of *Toll* is sufficient to drive cells along a ventral differentiation pathway. Double mutant combinations of dominant *Toll* alleles with recessive dorsalizing alleles at other loci have formed the basis of a pathway of action among these genes (*Figure 6*) (62; K.V.Anderson, unpublished results). Females carrying both *Toll*[D] and mutations in *dorsal* or *tube* produce dorsalized embryos: therefore even an overly active *Toll* product requires the activity of *dorsal* and *tube* to affect embryonic pattern, indicating that *dorsal* and *tube* act downstream of *Toll*. In contrast, females that are doubly mutant for the same $Toll^D$ allele and dorsalizing alleles of *gastrulation defective, nudel, pipe, snake* or *easter*, produce 'lateralized' embryos, in which all embryonic cells make lateral structures. Although this phenotype is not identical to that of the $Toll^D$ allele alone, the production of lateral structures in these embryos shows that none of these five gene products is absolutely required for the production of the morphogen. The results suggest that these five genes act upstream of *Toll* and are necessary for the activity of the wild-type *Toll* allele, but not the dominant allele. The phenotypes of double mutants of the ventralizing alleles of *cactus* and dorsalizing alleles in other loci are consistent with this ordering, and place *cactus* downstream of *Toll*, and place *dorsal* downstream of *cactus*, at the end of the pathway (K.V.Anderson, unpublished results).

The *dorsal* gene was cloned by chromosomal walking and jumping from a nearby cloned region (63). The product of the *dorsal* gene is a nuclear protein with strong sequence similarity to the avian oncogene *rel* (64,65). Since the genetic studies had indicated that maternal dorsal–ventral information is graded and that *dorsal* acts at the end of the pathway, it was satisfying to find, using antibodies directed against a bacterial *dorsal* fusion protein, that the *dorsal* protein is distributed in a gradient, with its highest concentration ventrally (65). The concentration of *dorsal* protein decreases gradually in lateral nuclei and no protein is detected in nuclei on the dorsal side of the embryo. The *dorsal* nuclear protein presumably influences, directly or indirectly, the transcription of a set of zygotic genes. There are probably several thresholds and several targets for *dorsal*. In fact, it seems likely that *dorsal* may act both as an activator and as a repressor, activating at least one zygotic gene (*snail* or *twist*) ventrally (66), and repressing at least two zygotic genes (*decapentaplegic* and *zerknüllt*) ventrally and laterally (67,68). It has been shown molecularly that *dorsal* is required for the localized expression of these zygotic genes in the syncytial blastoderm embryo, and biochemical experiments to test whether *dorsal* directly regulates their transcription can now be carried out. Although the *dorsal* protein is graded, the *dorsal* mRNA is uniformly distributed in the embryo (65). Therefore the mechanism that establishes the *dorsal* protein gradient must be different from the mechanism that establishes the *bicoid* gradient. Although the graded translation or stability of the *dorsal* protein must be a response to an already existing dorsal–ventral asymmetry in the embryo, none of the gene products that act upstream of *dorsal* seems to be pre-localized, thereby dictating the graded appearance of dorsal protein, Instead, protein–protein interactions among homogeneously distributed precursors appear to generate asymmetry after fertilization.

Phenotypic rescue experiments have indicated that the activity of the *Toll* gene product is central in generating dorsal – ventral polarity in the embryo (60,69). For mutants in each of the seven rescuable dorsal group loci, the rescuing activity appears to be the wild-type product of the gene, since cytoplasm from mutants in one locus can rescue mutants in all other loci as effectively as wild-type cytoplasm. Thus, in contrast to the anterior and posterior group genes, each of these gene products is independently synthesized and stored in the egg and the bioassay reflects the properties of the individual gene product. For mutants in all loci except *Toll*, the most ventral structure produced in the rescued embryo develops from the cells that lie on the curved, presumptive-ventral side of the egg, irrespective of the site where the cytoplasm was injected (60). Thus each of these mutants retains an underlying dorsal – ventral polarity. In contrast, embryos produced by $Toll^-$ mothers have no detectable dorsal – ventral asymmetry. Injection of $Toll^+$ gene product at any site in these embryos results in the production of ventral structures at that site, independent of the shape of the egg (69). For instance, if the purified transcript of the *Toll* gene is injected on the flat presumptive dorsal side of a $Toll^-$ embryo, mesoderm invaginates near the dorsal side of the eggshell rather than ventrally and the entire dorsal – ventral pattern is orientated with respect to the injection site. In these injected embryos, polarity is defined by the local high concentration of *Toll* product.

These rescue results indicate that in the wild-type embryo the local activity of the *Toll* product defines the ventral-most position in the embryo. However, molecular data indicate that both the *Toll* transcript and the *Toll* protein are uniformly distributed in the embryo (70; C.Hashimoto and K.V.Anderson, unpublished results). A plausible resolution to this paradox is that the *Toll* protein, although present uniformly throughout the dorsal – ventral axis, is normally active only on the ventral side, in response to some upstream asymmetric cue. In the rescue experiments, the requirement for the asymmetric cue is bypassed. Thus although some asymmetric signal must be generated during oogenesis, perhaps as a result of the activities of the *K10*, *gurken* and *torpedo* genes, the gene products that are active after fertilization are capable of organizing a complete normal pattern with another polarity.

Hints about what might constitute the active form of *Toll* and how that active form is confined to one part of the embryo have come from the molecular analysis of the *Toll* gene and of two genes that act upstream of *Toll*, *snake* and *easter*. DNA encoding *snake* was isolated from an existing chromosomal walk and shown to encode *snake* by transformation (71). The *Toll* and *easter* genes were identified molecularly by P-element transposon tagging (72; R.Chasan and K.V.Anderson, submitted). The *Toll* DNA sequence predicts that it encodes a transmembrane protein, with an extracytoplasmic domain of 800 amino acids and a cytoplasmic domain of 250 amino acids (72). The membrane localization of the *Toll* protein has been confirmed with antibodies directed against *Toll*, which recognize a cell surface protein (C.Hashimoto and K.V.Anderson, unpublished results). Both *snake* and *easter* appear to encode extracellular-type serine proteases (71; R.Chasan and K.V.Anderson, submitted). The path of information flow derived from double mutant studies indicates that information flows from *easter* and *snake* through *Toll* and ultimately to the nuclear *dorsal* protein, or, in terms of cellular locations, from an extracytoplasmic compartment, across the membrane through *Toll* to the cytoplasm, and then to the nucleus. These results raise the possibility that the *Toll* protein functions as a signal transducer. A local signal could cause the extracellular domain of the *Toll* protein to undergo a conformational change, which would then be transduced across the plasma membrane to alter the activity of the cytoplasmic domain of the *Toll* protein. If information flows from the *Toll* cytoplasmic domain to the *dorsal* protein, it presumably proceeds through one or more intermediates. Both *cactus* and *tube* act downstream of *Toll* but upstream of *dorsal*, and could act at this point in the pathway.

The ability of *Toll* to organize a pattern with any polarity in injection experiments suggests that *Toll* can also generate and amplify a 'ventral' signal in the absence of any external input. In addition, the interactions between *Toll* alleles have suggested that the *Toll* protein is a component of a positive autocatalytic loop (62). One scenario which incorporates all these features of *Toll* activity is that *Toll* acts as both a receptor and its own ligand in a self-amplifying cycle. That is, the extracytoplasmic domain of the *Toll* protein, together with other proteins in the extracytoplasmic compartment, can generate and amplify a signal which is then communicated to the cytoplasmic domain of the *Toll* protein. In normal development, an asymmetric trigger causes the local activation of a small amount of *Toll* on the ventral side of the embryo, and that signal is amplified locally by protein – protein interactions in an extracytoplasmic compartment, activating more *Toll* protein locally. In the injections into *Toll*⁻ embryos, the normal trigger is by-passed by a random activation of a small amount of *Toll* at the injection site, which then sets into motion the normal cascade of events including local amplification of the ventral signal.

Since both *snake* and *easter* appear to encode serine proteases that act upstream of *Toll*, proteolytic cleavages in an extracytoplasmic compartment are necessary to allow *Toll* activity. It is possible that proteolytic processing events are necessary to produce a cofactor that allows *Toll* to be active, or that *Toll* itself is the substrate for one or both proteases. There is genetic evidence that the level of activity of the *easter* protease is directly reflected in the pattern. Dominant *easter* alleles that ventralize or lateralize the pattern have been recovered at high frequency (R.Chasan and K.V.Anderson, submitted). Since *easter* acts upstream of *Toll*, such mutants in *easter* presumably exert their effect on the embryonic pattern by increasing the activity of *Toll*, directly or indirectly. Thus, although *Toll* need not be *easter*'s substrate, the proteolytic cleavage catalyzed by *easter* determines the level of activity of *Toll*.

The data on *Toll, easter* and *snake* suggest that the asymmetry of the dorsal – ventral pattern is defined by an initial activation of the *Toll* protein in an extracytoplasmic compartment followed by amplification of that activated state. With the available data, it is possible to speculate on how *Toll* might become activated and how that activation could be amplified, but in order to generate a pattern, that activation must be spatially regulated. A central question for future research is to characterize the nature of this spatial regulation.

5. Patterning processes with both maternal and zygotic contributions

In contrast to the processes of axis definition, which appear to require only maternally synthesized components, several early embryonic patterning processes require the activities of both maternal gene products and the products of embryonic transcription. Depending on the relative quantitative contributions of maternal and zygotic products, mutants in these processes may have been identified as maternal-effect or as zygotic lethal mutations.

Neurogenesis is a clear example of an early embryonic patterning process in which both maternal and zygotic contributions are essential. The nervous system of *Drosophila* originates by delamination of individual cells from the neurogenic ectoderm, during the extended germ band stage, 4 – 9 h after fertilization (73). Normally, 25% of the cells in the neurogenic ectoderm separate from the epithelium and contribute to the central nervous system, while the remaining cells from the same region develop as the ventral epidermis. When a cell delaminates to join the nervous system, it appears to signal its immediate neighbors in the neurogenic ectoderm, inhibiting them from also joining the nervous system (74). Mutations in at least nine neurogenic genes cause the entire neurogenic ectoderm to contribute to the nervous system, with a corresponding loss of the ventral epidermis (6,50,75). These mutations apparently disrupt the cell – cell signaling process by which neuroblasts inhibit neural development in their neighbors.

Most of the neurogenic genes are expressed both maternally and zygotically. The quantitative importance of the maternal and zygotic contributions varies among the genes, leading to the fact that some of the genes were identified on the basis of zygotic lethal mutations and others by maternal-effect mutations. The role of the *almondex* gene in neurogenesis, for instance, was defined on the basis of male-rescuable maternal-effect mutations (75). Individuals that are homozygous for *almondex* mutations can survive embryonic development if they have available the maternally inherited *almondex*⁺ product from their heterozygous mother. When such homozygous females reach the adult stage, they produce only mutant, neuralized, embryos if the zygotes also lack the *almondex* product. In this case, enough wild-type product to support development can be supplied either maternally or zygotically. At the other extreme is the *Notch* gene. Zygotes that lack *Notch* gene activity die as embryos, with a partially neuralized phenotype. It was possible to demonstrate the existence of a maternal supply of *Notch* gene product by generating germ line mosaics in which the germ cells lacked any *Notch* activity (76). These germ line cells could support normal embryonic development if the egg was fertilized by a wild-type sperm, but if fertilized by a *Notch*⁻ sperm, the neuralization in the resulting embryos (which lacked both the maternal and zygotic contribution) was more extreme than that observed in *Notch*⁻ embryos from heterozygous mothers, with all of the cells of the neurogenic ectoderm developing as nerve cells.

Similarly, most members of the *Polycomb* class of homeotic genes, which are globally required to maintain the expression of the segment-specific homeotic genes, appear to be expressed both maternally and zygotically (77,78). Loss-of-function alleles of a number of these genes are homozygous lethal, but embryos die without gross morphological defects. Extreme homeotic transformations of anterior segments towards more posterior identities are seen only when both components are removed by making germ line clones of the lethal mutations and fertilizing the eggs with mutant sperm.

Some genes required zygotically for segmentation must also be expressed maternally. Both maternal and zygotic expression of the *fused* gene are important for normal segmentation and expression of the segment polarity gene *armadillo* is necessary for normal oogenesis (79). However, most of the segmentation genes that subdivide the anterior – posterior pattern and the selector class of homeotic genes that give identities to specific segments are required only in the zygote. Why is it that some genes are expressed both maternally and zygotically while others are expressed exclusively zygotically? Conceivably those genes which must be expressed locally in a particular subset of cells are not expressed maternally because the mechanisms that localize maternal products are not sufficiently refined to ensure that only small subsets of cells acquire a particular gene product. If there is no special spatial or temporal constraint on the requirement for a particular gene product, then the existence of a maternal-effect or zygotic mutant phenotype reflects only how much gene product is needed before the critical phase of the developmental process, how much maternal product is stored, and the rate of zygotic transcription.

6. How many other genes are required maternally for embryonic pattern formation?

As is clear from the results described in this chapter, the identification and characterization of maternal-effect mutants that affect early embryonic development have yielded an enormous amount of information about the nature of pattern formation in the early *Drosophila* embryo. To understand pattern formation on a biochemical level it will be necessary to identify all the molecules that participate in the patterning pathways. Although the genetic screens for maternal-effect mutants have identified many key players in these processes, it is very likely that a number of important components have yet to be defined. The screens that have been carried out to identify maternal effect mutations have identified multiple alleles of the genes described in this chapter. However, Poisson statistics indicate that although most genes for which lack-of-function mutations cause a maternal-effect phenotype have been identified, a small number of such genes has probably been missed.

As is apparent from the preceding discussion of genes with both maternal and zygotic contributions to embryonic pattern formation, demonstrating a maternal function for a gene that is also required zygotically in embryonic development is a laborious process, requiring the construction of germ line mosaics. Therefore the number of genes that fall into this class is not yet clear, but will surely grow as more genes are cloned and maternal transcripts are identified.

Even more indecipherable is the number of truly pleiotropic genes that are required maternally for embryonic pattern formation and for other, unrelated, processes. For instance, if a gene product required for larval viability also has an important function maternally in embryonic development, it would not have been identified in the systematic screens for either zygotic or maternal-effect pattern mutants. As the genes identified on the basis of maternal-effect alleles are studied in more detail, an increasing number of them appear to be expressed and required at other phases of the life cycle. For instance, among the genes required maternally for dorsal – ventral pattern formation, genetic evidence suggests that at least *Toll, easter, cactus, tube* and *pelle* are also expressed zygotically (70; K.Anderson, S.Wasserman and P.Hecht, unpublished data). The *Toll* gene is required for larval viability, and the larvae that lack *Toll* die without any gross morphological defects (70). The identification of the critical and specific maternal role of the *Toll* product in embryonic pattern formation was made possible by the isolation of several partial loss-of-function and dominant gain-of-function alleles, while null alleles cause zygotic semi-lethality and therefore were not identified in screens for maternal-effect mutants. If the quantitative requirements for a gene product were such that partial loss-of-function alleles were lethal, then it would not be possible to isolate maternal-effect alleles. In a sample of germ line clones of random lethal mutations, 10% had a maternal effect on embryonic development (10), so it could be that more genes remain to be identified than have already been described. In the same sample of germ line clones, however, no new classes of phenotypes were identified. Therefore it seems likely that all the maternally dependent patterning pathways have been identified, even though not all the components have been identified. Certainly mutations in new maternally acting genes that are important for embryonic development will continue to be identified using classical genetic techniques, although at a slower pace than in the last 10 years. As the techniques of molecular biology become increasingly powerful, new genes will probably also first be identified on the basis of a biochemical interaction with the already identified components. Classical genetic analysis will then follow biochemistry to define the functions and relationships among these molecules.

7. References

1. Boveri,T. (1902) On multipolar mitosis as a means of analysis of the cell nucleus. Quoted in *Foundations of Experimental Embryology*. (1964) Willier,B.H. and Oppenheimer, J.M. (eds), Hafner Press, New York, pp. 74–97.
2. Nüsslein-Volhard,C. and Wieschaus,E. (1980) Mutations affecting segment number and polarity in *Drosophila*. *Nature*, **287**, 795–801.
3. Gans,M., Audit,C. and Masson,M. (1975) Isolation and characterization of sex-linked female sterile mutants in *Drosophila melanogaster*. *Genetics*, **81**, 683–704.
4. Mohler,J.D. (1977) Developmental genetics of the *Drosophila* egg. I. Identification of 50 sex-linked cistrons with maternal effect on embryonic development. *Genetics*, **85**, 259–272.
5. Mohler,J.D. and Carroll,A. (1984) Sex-linked female sterile mutations. *Drosophila Inf. Serv.*, **60**, 236–241.
6. Schüpbach,T. and Wieschaus,E. (1989) Female sterile mutations on the second chromosome of *Drosophila melanogaster*. I. Maternal effect mutations. *Genetics*, **121**, 101–117.
7. Spradling,A.C. and Rubin,G.M. (1982) Transposition of cloned P elements into *Drosophila* germ line chromosomes. *Science*, **218**, 341–347.
8. Mahowald,A.P. and Kambysellis,M.P. (1980) Oogenesis. In *The Genetics and Biology of Drosophila*. Ashburner,M. and Wright,T.R.F. (eds), Academic Press, New York, Volume 2d, pp. 141–224.
9. García-Bellido,A. and Robbins,L.G. (1983) Viability of female germ line cells homozygous for zygotic lethals in *Drosophila melanogaster*. *Genetics*, **103**, 235.
10. Perrimon,N., Engstrom,L. and Mahowald,A.P. (1984) Analysis of the effects of zygotic lethal mutations on the germ line functions in *Drosophila*. *Dev. Biol.*, **105**, 404–414.
11. Hough-Evans,B.R., Jacobs-Lorena,M., Cummings,M.R., Britten,R.J. and Davidson, E.H. (1980) Complexity of RNA in eggs of *Drosophila melanogaster* and *Musca domestica*. *Genetics*, **95**, 81–94.
12. Edgar,B.A. and Schubiger,G. (1986) Parameters controlling transcriptional activation during early *Drosophila* development. *Cell*, **44**, 871–877.
13. Spradling,A. (1981) The organization and amplification of two clusters of *Drosophila* chorion genes. *Cell*, **27**, 589–598.
14. Parks,S. and Spradling,A. (1987) Spatially regulated expression of chorion genes during *Drosophila* oogenesis. *Genes Dev.*, **1**, 497–509.

15. Anderson,K.V. and Lengyel,J.A. (1980) Changing rates of histone mRNA synthesis and turnover in *Drosophila* embryos. *Cell*, **21**, 717–727.
16. Anderson,K.V. and Lengyel,J.A. (1984) Histone gene expression in *Drosophila* development: multiple levels of gene regulation. In *Histone Genes, Structure, Organization and Regulation*. Stein,G.S., Stein,J.L. and Marzluff,W.F. (eds), Wiley, New York, pp. 135–161.
17. Natzle,J.E. and McCarthy,B.J. (1984) Regulation of alpha and beta tubulin genes during development. *Dev. Biol.*, **104**, 187–198.
18. Sodja,A., Arking,R. and Zafar,R.S. (1982) Actin gene expression during embryogenesis of *Drosophila melanogaster*. *Dev. Biol.*, **90**, 363–368.
19. Zalokar,M., Audit,C. and Erk,I. (1975) Developmental defects of female-sterile mutants of *Drosophila melanogaster*. *Dev. Biol.*, **47**, 419–432.
20. Freeman,M., Nüsslein-Volhard,C. and Glover,D.M. (1986) The dissociation of nuclear and centrosomal division in *gnu*, a mutation causing giant nuclei in *Drosophila*. *Cell*, **46**, 457–468.
21. Rice,T.B. and Garen,A. (1975) Localized defects of blastoderm formation in maternal effect mutants of *Drosophila*. *Dev. Biol.*, **43**, 277–286.
22. Frohnhöfer,H.G. and Nüsslein-Volhard,C. (1987) Maternal genes required for the anterior localization of *bicoid* activity in the embryo of *Drosophila*. *Genes Dev.*, **1**, 880–890.
23. Lohs-Schardin,M., Cremer,C. and Nüsslein-Volhard,C. (1979) A fate map for the larval epidermis of *Drosophila melanogaster*: localized cuticle defects following irradiation of the blastoderm with an ultraviolet laser microbeam. *Dev. Biol.*, **73**, 239–255.
24. Nüsslein-Volhard,C. (1979) Maternal effect mutations that alter the spatial coordinates of the embryo of *Drosophila melanogaster*. *Symp. Soc. Dev. Biol.*, **37**, 185–211.
25. Foe,V.E. (1988) Mitotic domains reveal early commitment of cells in *Drosophila* embryos. *Development*, in press.
26. Nüsslein-Volhard,C., Frohnhöfer,H.G. and Lehmann,R. (1987) Determination of anteroposterior polarity in *Drosophila*. *Science*, **238**, 1675–1681.
27. Anderson,K.V. (1987) Dorsal–ventral embryonic pattern genes of *Drosophila*. *Trends Genet.*, **3**, 91–97.
28. Bull,A. (1966) *Bicaudal*, a genetic factor which affects the polarity of the embryo of *Drosophila melanogaster*. *J. Exp. Zool.*, **161**, 221–242.
29. Nüsslein-Volhard,C. (1977) Genetic analysis of pattern formation in the embryo of *Drosophila melanogaster*: characterization of the maternal-effect mutant *bicaudal*. *Roux's Arch. Dev. Biol.*, **183**, 249–268.
30. Mohler,J. and Wieschaus,E.F. (1986) Dominant maternal-effect mutations of *Drosophila melanogaster* causing the production of double-abdomen embryos. *Genetics*, **112**, 803–822.
31. Lohs-Schardin,M. (1982) *Dicephalic*—a *Drosophila* mutant affecting polarity in follicle organization and embryonic patterning. *Roux's Arch. Dev. Biol.*, **191**, 28–36.
32. Frohnhöfer,H.G. and Nüsslein-Volhard,C. (1986) Organization of anterior pattern of the *Drosophila* embryo by the maternal gene *bicoid*. *Nature*, **324**, 120–125.
33. Schüpbach,T. and Wieschaus,E. (1986) Maternal-effect mutations altering the anterior–posterior pattern of the *Drosophila* embryo. *Roux's Arch. Dev. Biol.* **195**, 302–317.
34. Frigerio,G., Burri,M., Bopp,D., Baumgartner,S. and Noll,M. (1986) Structure of the segmentation gene *paired* and the *Drosophila* PRD gene set as part of a gene network. *Cell*, **47**, 735–746.
35. Berleth,T., Burri,M., Thoma,G., Bopp,D., Richstein,S., Frigerio,G., Noll,M. and Nüsslein-Volhard,C. (1988) The role of localization of *bicoid* RNA in organizing the anterior pattern of the *Drosophila* embryo. *EMBO J.*, **7**, 1749–1756.
36. Driever,W. and Nüsslein-Volhard,C. (1988) A gradient of *bicoid* protein in *Drosophila* embryos. *Cell*, **54**, 83–93.
37. Driever,W. and Nüsslein-Volhard,C. (1988) The *bicoid* protein determines position in the *Drosophila* embryo in a concentration-dependent manner. *Cell*, **54**, 95–104.
38. Carroll,S.B., Winslow,G.M., Schüpbach,T. and Scott,M.P. (1986) Maternal control of *Drosophila* segmentation gene expression. *Nature*, **323**, 779–789.
39. Mlodzik,M., De Montrion,C.M., Hiromi,Y., Krause,H.M. and Gehring,W.J. (1987) The influence on the blastoderm fate map of maternal-effect genes that affect the antero-posterior pattern in *Drosophila*. *Genes Dev.*, **1**, 603–614.

40. Tautz,D. (1988) Regulation of the *Drosophila* segmentation gene *hunchback* by two maternal morphogenetic centers. *Nature*, **332**, 281–284.
41. Mlodzik,M. and Gehring,W.J. (1987) Hierarchy of the genetic interactions that specify the anteroposterior segmentation pattern of the *Drosophila* embryo as monitored by *caudal* protein expression. *Development*, **101**, 421–435.
42. Macdonald,P.M. and Struhl,G. (1986) A molecular gradient in early *Drosophila* embryos and its role in specifying the body pattern. *Nature*, **324**, 537–545.
43. Stephenson,E.C. and Mahowald,A.P. (1987) Isolation of *Drosophila* clones encoding maternally restricted RNAs. *Dev. Biol.*, **124**, 1–8.
44. Lehmann,R. and Nüsslein-Volhard,C. (1986) Abdominal segmentation, pole cell formation and embryonic polarity require the localized activity of *oskar*, a maternal gene in *Drosophila*. *Cell*, **47**, 141–152.
45. Illmensee,K. and Mahowald,A.P. (1974) Transplantation of posterior pole plasm in *Drosophila*. Induction of germ cells at the anterior pole of the egg. *Proc. Natl. Acad. Sci. USA*, **71**, 1016–1020.
46. Lehmann,R. (1985) PhD thesis, Universität Tübingen.
47. Lasko,P.F. and Ashburner,M. (1988) The product of the *Drosophila* gene *vasa* is very similar to eukaryotic initiation factor 4A. *Nature*, **335**, 611–617.
48. Lehmann,R. and Nüsslein-Volhard,C. (1987) Involvement of the *pumilio* gene in the transport of an abdominal signal in the *Drosophila* embryo. *Nature*, **329**, 167–170.
49. Frohnhöfer,H.G. (1987) PhD thesis, Universität Tübingen.
50. Perrimon,N., Mohler,D., Engstrom,L. and Mahowald,A.P. (1986) X-linked female-sterile loci in *Drosophila melanogaster*. *Genetics*, **113**, 695–712.
51. Degelmann,A., Hardy,P.A., Perrimon,N. and Mahowald,A.P. (1986) Developmental analysis of the torso-like phenotype in *Drosophila* produced by a maternal-effect locus. *Dev. Biol.*, **115**, 479–489.
52. Klingler,M., Erdélyi,M., Szabad,J. and Nüsslein-Volhard,C. (1988) Function of *torso* in determining the terminal anlagen of the *Drosophila* embryo. *Nature*, **335**, 275–277.
53. Strecker,T.A., Merriam,J.R. and Lengyel,J.A. (1988) Graded requirement for the zygotic terminal gene, *tailless*, in the brain and tail regions of the *Drosophila* embryo. *Development*, **102**, 721–734.
54. Wieschaus,E. (1979) fs(1)*K10*, a female sterile mutation altering the pattern of both the egg coverings and the resultant embryos in *Drosophila*. In *Cell Lineage, Stem Cells and Cell Differentiation*. LeDouarin,N. (ed.), pp 291–302.
55. Schüpbach,T. (1987) Germ line and soma cooperate during oogenesis to establish the dorsoventral pattern of egg shell and embryo in *Drosophila melanogaster*. *Cell*, **49**, 699–707.
56. Wieschaus,E., Marsh,J.L. and Gehring,W.J. (1978) *fs(1)K10*, a germ-line dependent female-sterile mutation causing abnormal chorion morphology in *Drosophila melanogaster*. *Roux's Arch. Dev. Biol.*, **184**, 75–82.
57. Haenlin,M., Steller,H., Pirrotta,V. and Mohier,E. (1985) A 43 kilobase cosmid P transposon rescues the fs(1)*K10* morphogenetic locus and three adjacent *Drosophila* developmental mutants. *Cell*, **40**, 827–837.
58. Haenlin,M., Roos,C., Cassab,A. and Mohier,E. (1987) Oocyte-specific transcription of fs(1)*K10*: a *Drosophila* gene affecting dorsal–ventral developmental polarity. *EMBO J.*, **6**, 801–807.
59. Prost,E., Deryckere,F., Roos,C., Haenlin,M., Pantesco,V. and Mohier,E. (1988) Role of the oocyte nucleus in determination of the dorsoventral polarity of *Drosophila* as revealed by molecular analysis of the *K10* gene. *Gene Dev.*, **2**, 891–900.
60. Anderson,K.V. and Nüsslein-Volhard,C. (1986) Dorsal-group genes of *Drosophila*. *Symp. Soc. Dev. Biol.*, **44**, 177–194.
61. Nüsslein-Volhard,C., Lohs-Schardin,M., Sander,K. and Cremer,C. (1980) A dorso-ventral shift of embryonic primordia in a new maternal-effect mutant of *Drosophila*. *Nature*, **283**, 474–476.
62. Anderson,K.V., Jürgens,G. and Nüsslein-Volhard,C. (1985) Establishment of dorsal-ventral polarity in the *Drosophila* embryo: genetic studies on the role of the *Toll* gene product. *Cell*, **42**, 779–789.
63. Steward,R., McNally,F.J. and Schedl,P. (1984) Isolation of the *dorsal* locus of *Drosophila*. *Nature*, **311**, 262–265.
64. Steward,R. (1987) *Dorsal*, an embryonic polarity gene in *Drosophila*, is homologous to the vertebrate proto-oncogene, c-*rel*. *Science*, **238**, 692–694.

65. Steward,R., Zusman,S., Huang,L. and Schedl,P. (1988) The *dorsal* protein is distributed in a gradient in early *Drosophila* embryos. *Cell*, in press.

66. Thisse,B., Stoetzel,C., El Messal,M. and Perrin-Schmitt,F. (1987) Genes of the *Drosophila* maternal dorsal group control the specific expression of the zygotic gene *twist* in presumptive mesodermal cells. *Genes Dev.*, **1**, 709–715.

67. Rushlow,C., Frasch,M., Doyle,H. and Levine,M. (1987) Maternal regulation of *zerknüllt*: a homeobox gene controlling differentiation of dorsal tissues in *Drosophila*. *Nature*, **330**, 583–586.

68. St. Johnston,R.D. and Gelbart,W.M. (1987) *Decapentaplegic* transcripts are localized along the dorsal–ventral axis of the *Drosophila* embryo. *EMBO J.*, **6**, 2785–2791.

69. Anderson,K.V., Bokla,L. and Nüsslein-Volhard,C. (1985) Establishment of dorsal–ventral polarity in the *Drosophila* embryo: the induction of polarity by the *Toll* gene product. *Cell*, **42**, 791–798.

70. Gerttula,S., Jin,Y. and Anderson,K.V. (1988) Zygotic expression and activity of the *Drosophila Toll* gene, a gene required maternally for embryonic dorsal–ventral pattern formation. *Genetics*, **119**, 123–133.

71. DeLotto,R. and Spierer,P. (1986) A gene required for the specification of dorsal–ventral pattern in *Drosophila* appears to encode a serine protease. *Nature*, **323**, 688–692.

72. Hashimoto,C., Hudson,K.L. and Anderson,K.V. (1988) The *Toll* gene of *Drosophila*, required for dorsal–ventral embryonic polarity, appears to encode a transmembrane protein. *Cell*, **52**, 269–279.

73. Hartenstein,V. and Campos-Ortega,J.A. (1984) Early neurogenesis in wild-type *Drosophila melanogaster*. *Roux's Arch. Dev. Biol.*, **193**, 308–325.

74. Doe,C.Q. and Goodman,C.S. (1985) Neurogenesis in grasshopper and *fushi tarazu* Drosophila embryos. *Cold Spring Harbor Symp. Quant. Biol.*, **50**, 891–903.

75. Lehmann,R., Jiménez,F., Dietrich,U. and Campos-Ortega,J.A. (1983) On the phenotype and development of mutants of early neurogenesis in *Drosophila melanogaster*. *Roux's Arch. Dev. Biol.*, **192**, 62–74.

76. Jiménez,F. and Campos-Ortega,J.A. (1982) Maternal effects of zygotic mutants affecting early neurogenesis in *Drosophila*. *Roux's Arch. Dev. Biol.*, **191**, 191–201.

77. Ingham,P.W. (1984) A gene that regulates the *biothorax* complex differentially in larval and adult cells of *Drosophila*. *Cell*, **37**, 815–823.

78. Breen,T.R. and Duncan,I. (1986) Maternal expression of genes that regulate the *Bithorax* complex of *Drosophila melanogaster*. *Dev. Biol.*, **118**, 442–456.

79. Wieschaus,E. and Noell,E. (1986) Specificity of embryonic lethal mutations in *Drosophila* analyzed in germ line clones. *Roux's Arch. Dev. Biol.*, **195**, 63–73.

80. Boswell,R. and Mahowald,A.P. (1985) *tudor*, a gene required for assembly of the germ plasm in *Drosophila melanogaster*. *Cell*, **43**, 97–104.

81. Konrad,K.D., Goralski,T.J. and Mahowald,A.P. (1988) Developmental analysis of fs(1) *gastrulation defective*, a dorsal-group gene of *Drosophila melanogaster*. *Roux's Arch. Dev. Biol.*, **197**, 75–91.

82. Müller-Holtkamp,F., Knipple,D.C., Seifert,E. and Jäckle,H. (1985) An early role of maternal mRNA in establishing the dorsoventral pattern in *pelle* mutant *Drosophila* embryos. *Dev. Biol.*, **110**, 238–246.

2

Drosophila: the zygotic contribution
Michael S.Levine and Katherine W.Harding

1. Introduction

In *Drosophila*, embryonic cells come to express distinct sets of genes and follow different pathways of morphogenesis based on their spatial coordinates within the early embryo. Many of the genes that specify this positional information have been identified on the basis of mutations that disrupt cuticular structures secreted by the epidermis of mid-staged embryos. At least 40 zygotically active genes have been identified and these represent the vast majority of regulatory genes that control the embryonic pattern. These genes fall into two broad groups:
(i) those that differentiate the anterior – posterior pattern;
(ii) those that control the dorsal – ventral pattern.

Nearly two-thirds of the known anterior – posterior pattern genes have been cloned and characterized, providing considerable information about the proteins they encode and the basis for their stringently regulated patterns of expression during development. Most of these genes encode proteins that possess either the homeobox or zinc finger DNA-binding motif (see *Table 1*), suggesting that they control development by modulating gene expression at the level of transcription. Furthermore, nearly every one of these genes shows a unique pattern of expression, suggesting that many or all of the 6000 cells that comprise the early embryo contain a unique combination of regulatory gene products. It is thought that these different permutations of gene expression play an important role in selecting cell fate and specifying diverse patterns of morphogenesis. The abnormal expression of one or more of these genes in the early embryo can result in transformations of cell fate and subsequent disruptions in the spatial organization of advanced-stage embryos. Here we consider possible mechanisms underlying the precise expression of these key regulatory genes in the early embryo.

Table 1. Zygotic genes required for segmentation in *Drosophila*

	Cloned?	Nuclear?	DNA-binding motif?	References
Gap genes				
1. *hunchback*	yes	yes	fingers[a]	47
2. *Krüppel*	yes	yes	fingers	46,67
3. *knirps*	no	?	?	
4. *tailless*	no	?	?	
5. *giant*	no	?	?	
Pair-rule genes				
1. *fushi tarazu*	yes	yes	homeobox[b]	50,51
2. *even-skipped*	yes	yes	homeobox	53,54,118
3. *paired*	yes	yes	homeobox	55
4. *hairy*	yes	yes	no	52
5. *runt*	yes	?	?	56
6. *odd-skipped*	no	?	?	
7. *odd-paired*	no	?	?	
8. *sloppy-paired*	no	?	?	
Segment polarity genes				
1. *engrailed*	yes	yes	homeobox	58,138
2. *gooseberry*	yes	?	homeobox	60,61
3. *wingless*	yes	no	no	59
4. *patched*	no	?	?	
5. *hedgehog*	no	?	?	
6. *armadillo*	no	?	?	
7. *ci*[D]	no	?	?	
8. *fused*	no	?	?	
9. *naked*	no	?	?	
Homeotic genes				
1. *Labial*	yes	yes	homeobox	24,25
2. *proboscipedia*	yes	yes	homeobox	23
3. *Deformed*	yes	yes	homeobox	139
4. *Sex combs reduced*	yes	yes	homeobox	30
5. *Antennapedia*	yes	yes	homeobox	28,29
6. *Ultrabithorax*	yes	yes	homeobox	32
7. *abdominal-A*	yes	yes	homeobox	33,34
8. *Abdominal-B*	yes	yes	homeobox	33
9. *caudal*	yes	yes	homeobox	15,16,24

[a]Zinc finger proteins all contain a DNA-binding domain of a type originally described in TFIIIA, a *Xenopus* RNA polymerase III transcriptional factor that regulates the expression of 5S RNA genes during development. The DNA-binding domain usually comprises a number of repeats of 27–28 amino acids with a characteristic distribution of cysteine, phenylalanine, leucine and histidine residues which can chelate zinc ions. Structural features of zinc finger domains have been recently reviewed (141). This large family of proteins is implicated in the developmental regulation of gene expression.
[b]The homeobox is a segment of DNA encoding a 60 amino acid domain found within many proteins. It was first identified as a region of homology common to several homeotic genes of *Drosophila* and to the segmentation gene *ftz* (34,35). Homeobox-containing genes have now been found in many species including *Caenorhabditis*, *Xenopus* and the mouse (see other chapters in this book). Many of these genes are implicated as playing a role during development. Homeoboxes have recently been demonstrated in several mammalian transcription factors, and it seems likely that proteins containing such sequences are DNA-binding proteins with this broader role. These properties of homeobox proteins have been recently reviewed (140).

2. The regulatory hierarchies

2.1 Saturation mutagenesis for disruption of the embryonic cuticle

During the mid-point of embryogenesis (about 12 h after fertilization) the epidermis secretes a cuticle that contains a number of simple visual markers for segmentation (1) (see *Figure 1*). Genetic screens which attempt to achieve saturation mutagenesis have been carried out in an effort to identify all of the genes in the *Drosophila* genome that disrupt these cuticular pattern elements (2 – 12). The vast majority of mutants that cause embryonic lethality do not cause specific pattern defects. Mutations in only about 3% of the known embryonic lethal complementation groups in *Drosophila* disrupt the cuticular pattern. Approximately 30 of these loci are specifically required for the

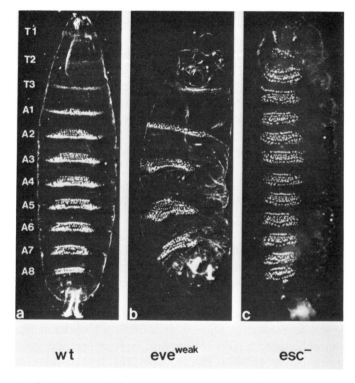

Figure 1. Cuticles of wild-type and mutant embryos. The embryos shown are orientated so that anterior is up and the ventral surface is displayed. (**a**) Wild-type (wt) embryo. The eight abdominal (A1 – A8) and three thoracic (T1 – T3) denticle belts can be seen. (**b**) Embryo homozygous for a weak *eve* mutant allele. This embryo shows a typical pair-rule phenotype, having only one-half the normal number of segments due to the deletion of alternating segments. (**c**) Embryo showing the *esc⁻* phenotype. All segments have been homeotically transformed towards the eighth abdominal segment.

morphogenesis of the segmentation pattern (5 – 7). Examples of mutations that alter the segmentation pattern are shown in *Figure 1*.

   Genetic screens already carried out have probably identified nearly every gene that controls the differentiation of the anterior – posterior (A/P) pattern in middle body regions of the early embryo. However, it is possible that not all of the genes acting on the anterior and posterior poles have been identified. The structures that occur near the anterior pole are obscured by the process of head involution, whereby the 4 – 6 segments that comprise the head are invaginated and not easily visualized (13). Similar problems pertain to the identification of genes controlling the morphogenesis of posterior structures, as well as internal structures such as the gut. In fact, proof that not all of the genes were identified in the past mutagenesis screens has been provided by the isolation of additional genes using molecular probes. For example, a gene called *caudal (cad)* was not identified in the original screens, but instead was isolated on the basis of cross-homology with a homeobox-encoding DNA fragment (see footnote to *Table 1*) (14 – 16). Cloned *cad* sequences were used to identify the cytogenetic map position of this gene by *in situ* hybridization to polytene chromosomes. This in turn provided the basis for obtaining mutations in the gene (15). *cad* mutations cause a disruption in the morphogenesis of the telson, which is a structure derived from the posterior-most regions of the embryo. The discovery of *cad* is an important reminder that in considering regulatory interactions among segmentation genes probably not all of the relevant genes have yet been identified. It is therefore likely that our current view of the segmentation regulatory network is inadequate.

2.2   Classification of the genes that control the segmentation pattern

Approximately 30 different zygotically active genes are required for the differentiation of the embryonic A/P pattern and the specification of morphologically diverse body segments (summarized in *Table 1*). These genes fall into two groups: segmentation genes and homeotic genes. Segmentation genes subdivide the early embryo into a repeating series of homologous segment primordia (4). Homeotic genes act afterwards to diversify the segmentation pattern such that each segment acquires a distinct morphology (2,3,17). Mutations in segmentation genes alter segment number or polarity. For example, the segmentation mutation [*even-skipped (eve)*] shown in *Figure 1b* lacks alternating segments and therefore possesses only one-half of the normal number of segments (4). In contrast, mutations in homeotic genes do not disrupt segmentation *per se*, but instead cause transformations in segment identity such that a part or all of one segment takes on the characteristics of another segment (2,3). *Figure 1c* shows a rather dramatic homeotic transformation [*extra sex combs*[t] *esc*[−]], whereby all segments, including those that comprise the head, display a cuticular pattern characteristic of the normal eighth abdominal segment (A8). This particular phenotype (18) results from the abnormal expression of several homeotic genes in segments where they are normally not expressed (18–20).

There are two very different types of homeotic genes:
(i) homeotic selector genes, which directly influence segment morphogenesis;
(ii) homeotic regulatory genes, which influence morphogenesis indirectly by regulating the expression of the selector genes (21).

We will focus primarily on the selector class of homeotic genes in this review. There are at least nine such genes, as summarized in *Table 1* (2,3,22–26). Most of these occur within either the *Antennapedia* (*ANTP-C*) or *Bithorax* (*BX-C*) gene complex. The *ANTP-C* contains five of the genes (3,23–25,27–31), and the *BX-C* contains three (2,22,32,33). Each of these genes contains a homeobox (34,35) (see footnote b to *Table 1*), and the six proteins that have been examined were found to accumulate in the nucleus (23,36–43). It should be noted that there is some question as to what constitutes a homeotic gene in the *BX-C*. According to the Lewis model the *BX-C* contains at least eight homeotic genes (2). However, recent molecular and genetic studies suggest that there are only three embryonic lethal complementation groups within the *BX-C* (22,32,33). Many of the genetic elements that were originally classified as individual genes might correspond to *cis*-regulatory sequences that modulate the expression of the three primary genes.

The segmentation genes can be divided into three classes (*Table 1*).

(i)   Gap genes. The gap genes divide the embryo into broad regions that each encompass several segment primordia. Gap mutants show deletions of contiguous blocks of segments. For example, advanced-stage *knirps* (*kni*) mutant embryos lack the first through seventh abdominal segments, whereas the remaining segments appear normal (4).

(ii) Pair-rule genes. These genes subdivide the broad 'gap' domains into individual segments. Pair-rule mutations usually cause the deletion of pattern elements in alternating segments, as shown in *Figure 1b*.

(iii) Segment polarity genes. These genes are required for the morphogenesis of segmentally repeated structures. Mutations in this class of genes cause the loss of certain structures in each segment, and the duplication of different pattern elements (4). *engrailed* (*en*) is the best characterized segment polarity gene, and is required for the establishment and maintenance of segment boundaries (44,45). Progressively more severe *en* mutants cause the increasing loss of segment borders, and fusions between adjacent segments.

Unlike the homeotic genes, segmentation genes tend not to be clustered but are scattered throughout the genome. Two of the five gap genes have been cloned. Sequence analysis suggests that both encode proteins that contain multiple copies of a zinc finger DNA-binding motif (see footnote a to *Table 1*) (46,47). Immunolocalization studies show that both proteins accumulate in the nucleus of those cells where they are expressed (46,47). Five of the eight pair-rule genes have been cloned, and three of these contain a homeobox (50 – 56) (see *Table 1*). The *hairy* protein definitely does not contain a homeobox, or other known DNA-binding motif (C.Rushlow, personal communication), while the nature of the *runt* protein is currently unknown. Interestingly, recent immunolocalization studies indicate that the *hairy* protein accumulates in the nucleus, suggesting that it might exert a relatively direct influence on the control of gene expression (57). Finally, three of the nine segment polarity genes have been cloned, and two contain homeoboxes (58 – 61). The third gene, *wingless* (*wg*), encodes a protein that does not accumulate in the nucleus and is likely to control development through a cell – cell signal transduction mechanism (59). The *wg* protein shares extensive homology with the mouse *int-1* gene product, which has been shown to specify a diffusable growth factor (62). There is evidence that *wg* influences the segmentation pattern by influencing the activities of *en* and other segment polarity genes in neighboring cells (63,64).

2.3 A vast array of regulatory gene expression patterns

Each of the genes listed in *Table 1* shows a unique pattern of expression during early development. Many of the homeotic, gap and pair-rule genes begin to show spatially restricted patterns of expression prior to the completion of cellularization, at about 2 – 2.5 h after fertilization (see refs 65 – 67 for examples). The three segment polarity genes that have been examine are not selectively expressed until slightly later stages, after the other classes of segmentation genes show restricted patterns (60,61,68,69). *Figure 2* shows examples of the different classes of segmentation and homeotic gene expression patterns. In general, the first transcripts that are detected for each gene are rather broadly distributed, but rapidly become more tightly localized during cellularization. For example, transcripts encoded by the gap gene *Krüppel* (*Kr*) are first detected during cleavage cycle 10, and by cycle 11 they are observed as a broad band over the middle body region of the embryo (67). However, by the start of cell cycle 14, this band has narrowed and is restricted to a central region of the embryo corresponding to parasegments four through six (*Figure 2a*).

The term parasegment has been coined by Martinez-Arias and Lawrence (70) to describe the first morphological sign of segmentation, which appears as a series of thickenings in the embryonic mesoderm. This is followed by the appearance of a series of grooves in the outer surface in the ectoderm. These grooves divide the embryo into 14 parasegments corresponding to the domains of expression of many of the genes essential for segmentation and segment identity. The parasegments do not coincide to the segments seen in the larval or adult stages, but include cells in the

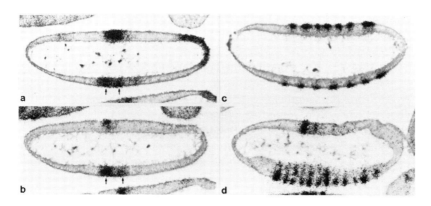

Figure 2. Patterns of segmentation and homeotic gene expression. All embryos are 2.5 – 3 h old and are oriented with the anterior pole to the left and dorsal side up. (**a**) Embryo hybridized with a *Kr* RNA probe. Strong expression is seen in the thorax (arrows) and at the posterior pole. (**b**) Serial section of the same embryo shown in panel (a) after hybridization with an *Antp* RNA probe. *Antp* expression is detected in the thorax, in almost the same sites as seen for *Kr* (arrows). (**c**) A different embryo hybridized with a *ftz* RNA probe. Seven bands of *ftz* expression can be seen, corresponding to the even-numbered parasegments. (**d**) Another embryo hybridized with an *en* RNA probe. *en* transcripts are detected in the posterior compartment of each segment.

posterior part of one segment and the anterior part of the next.

By the completion of cellular blastoderm formation each segment primordium contains a unique combination of gap, pair-rule and homeotic gene products. This corresponds to the stage when embryonic cells become irreversibly committed to specific pathways of segment morphogenesis. At this time, each segment primordium is composed of a strip, of about four cells in width, that encircles the embryo. More precise restrictions in cell fate occur at gastrulation, when each of the segment polarity genes is expressed within specific cells of the segment primordia. For example, the segment polarity gene *en* is expressed in the posterior-most cell of each primordium, where it is required for the establishment and maintenance of the posterior compartment (44,45,68,71) (see *Figure 2d*).

The precise patterns of regulatory gene expression that are established during early development are responsible for the segmentation pattern seen in advanced-stage embryos. The expression of one or more of these regulatory genes in cells where they are not normally expressed can cause severe disruptions of the pattern. For example, mutations in the homeotic regulatory genes *esc* (18,19) and *Polycomb* (*Pc*) (72) result in the abnormal expression of posterior homeotic selector genes, including *Abdominal-B* (*Abd-B*), in anterior segments where they effect homeotic transformations (20,73) (see *Figure 1c* and *5*). Ectopic expression of different segmentation genes and homeotic genes has also been brought about by reintroducing these genes into flies by P-element germ line transformation such that their expression is now under the control of the heat inducible hsp70 promoter (74,75). For example, the homeotic gene *Antennapedia* (*Antp*) is required for the morphogenesis of thoracic structures, particularly those derived from parasegment 4 (PS4) (76,77). *Antp* mutants display a phenotype whereby antennal structures are transformed to corresponding leg structures. As shown in *Figure 2c*, *Antp* transcripts are restricted to the presumptive thorax of early wild-type embryos, with highest levels of expression in PS4 and the metathorax (T3) (65). The hsp70 promoter has been used to drive ectopic expression of *Antp* products in regions where they are normally not detected, such as in the presumptive head (75). This ectopic expression of *Antp* can cause homeotic transformations, whereby head structures follow a pathway of morphogenesis characteristic for cells that reside within PS4. Depending on the time when mis-expression is induced, it is possible to obtain the 'classical' *Antp* transformation of antennal structures into leg structures.

2.4 Summary of the regulatory hierarchy

The precise expression of each segmentation and homeotic gene depends on a complex hierarchy of interactions, as summarized in *Figure 3*. This hierarchy culminates with the selective expression of homeotic and segment polarity genes in gastrulating embryos. This specifies the repeated pattern of morphologically diverse body segments. The regulatory relationships that have been established are largely based on examining the distribution of RNAs and/or proteins encoded by a given gene in embryos carrying mutations for other segmentation genes. The first step in the segmentation process appears to involve localized expression of the five gap genes in discrete domains along the A/P body axis. Recent studies have shown that maternal factors deposited into the egg during oogenesis play an essential role in the establishment of gap gene expression (78–80 and Chapter 1). The maternal gene *bicoid* (*bcd*) is of particular interest and importance in this regard (81). *bcd* products are distributed in a concentration gradient along the A/P axis, with peak levels at the anterior pole (78,79). High concentrations of *bcd* product define the initial limits of expression of the gap gene *hunchback* (*hb*), thereby restricting the primary domain of *hb* function to anterior regions of the early embryo (82). Additional maternal factors, acting together with *hb*, help establish the patterns of *Kr* and *kni* expression in more posterior regions of the embryo (83).

The gap genes influence the segmentation process indirectly by regulating the expression of pair-rule and homeotic genes (84–88). One of the central problems of the segmentation process is how relatively few gap genes provide sufficient information to define highly localized 'stripes' of pair-rule and homeotic gene expression. According to a model proposed

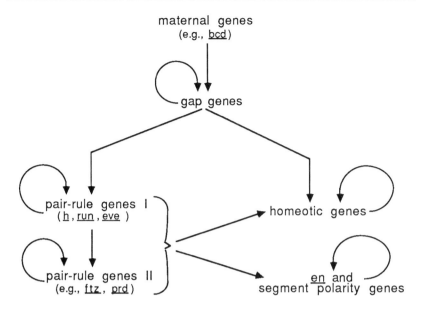

Figure 3. Summary of the segmentation gene hierarchy. Straight arrows indicate the regulation of one class of genes by another. Curved arrows indicate cross-regulation among genes of the same class. The initial broad domains of gap gene expression are defined by the maternal effect genes. Cross-regulation among the gap genes is required for the refinement and maintenance of expression within discrete regions along the A/P axis of early embryos. The unique expression patterns of both the homeotic genes and the early pair-rule (pair-rule genes I) appear to be established through differential responses to combinations of gap gene products. The maintenance of these patterns requires cross-regulatory interactions. The early pair-rule genes initiate the periodic expression patterns of the late pair-rule genes (pair-rule genes II). Cross-regulation among the late pair-rule genes is needed for the maintenance of their patterns. Both classes of pair-rule genes appear to be involved in the modulation and refinement of the homeotic expression patterns. In addition, pair-rule genes are required for the initiation of *en* and other segment polarity genes. Cross-regulation among *en* and the segment polarity genes is subsequently required to maintain and modify these patterns.

by Meinhardt, boundaries between adjacent gap gene products provide the basis for these sharp stripes of gene expression (89). That is a particular combination of gap gene products might selectively activate or repress the expression of a given target gene. The pair-rule genes, in turn, implement segmentation by modulating the expression of homeotic genes (90), and through the regulation of *en* and other segment polarity genes (53,54,69,91).

   In addition to the maternal/gap/pair-rule/homeotic/segment polarity gene hierarchy shown in *Figure 3*, regulatory interactions have also been shown to occur within each class of genes. For example, regulatory interactions between gap genes are important for delineating the precise limits of their expression (92) and some pair-rule genes (such as *hairy* and *runt*) play critical roles in defining the periodic expression of others (i.e. *ftz*) (52,85,91). Autoregulation has been shown to be important for sustaining high levels of homeotic gene products within certain embryonic tissues (93) and has also been implicated in the refinement of *ftz* and *eve* expression stripes during gastrulation (94,95).

Many of the regulatory interactions shown in *Figure 3* are likely to occur at the level of transcription since the majority of these genes encode proteins that contain either the homeobox or zinc finger DNA-binding motif (summarized in *Table 1*). The *bcd* protein also contains a homeobox, suggesting that the first step in the segmentation hierarchy also occurs at the level of transcription (96). Here we will review several model regulatory interactions that illustrate each of the steps shown in the hierarchy. In particular, cross-regulatory interactions among *ANTP-C* and *BX-C* homeotic selector genes provide a model for lateral interactions among members of a single class of genes. The control of *en* expression by the pair-rule genes *eve* and *ftz* serves as a model for the establishment of segment polarity gene expression. *eve* and *ftz* autoregulation might provide general clues regarding the role of autoregulation in the control of the segmentation regulatory network. Finally, we will consider how gap genes influence the expression of several pair-rule and homeotic genes, including *eve* and *Antp*.

3.   Maintenance of homeotic expression patterns

3.1   Correlation between sites of *Antp* expression and function

Eight of the nine known homeotic selector genes reside within the *BX-C* or *ANTP-C* (*Table 1*). Genes of the *BX-C* are required for the specification of segment identities in posterior regions of the fly, including the mesothorax through the eighth abdominal segment (2,22,33,97). Embryos that lack all *BX-C* functions show a transformation of the posterior thoracic and abdominal segments towards a more anterior fate. In contrast, *ANTP-C* homeotic genes are required for the specification of anterior body segments, including the anterior thorax and most or all of the head segments (3,76). As mentioned above, the *BX-C* contains three homeotic-lethal complementation groups (22). The *ANTP-C* is somewhat more complex since it includes at least five homeotic selector genes, as well as the maternal gene *bcd*, the dorsal–ventral polarity gene *zen* (23,26,98,99) and the pair-rule gene *ftz*. Each of the eight *ANTP-C/BX-C* homeotic selector genes, as well as *cad*, contains a homeobox. It has been proposed that an ancestral homeotic gene complex might have contained all of the *ANTP-C* and *BX-C* genes, as well as *cad* (16).

Cloned DNAs have been used to localize transcripts encoded by each of these genes in tissue sections of wild-type embryos (reviewed in ref. 100). Furthermore proteins encoded by six of the nine homeotic selector genes have been localized using specific antibodies (36–43). The regions of wild-type embryos that contain products encoded by each of these loci correspond quite closely with the embryonic segments that are most disrupted in mutants for these genes. This suggests that the establishment of localized domains of homeotic gene function is largely a function of transcriptional control. However, it should be noted that to date, the examination of expression patterns has been performed by analyzing steady-state levels of RNAs. It is conceivable that selective patterns of transcript distribution result from a post-transcriptional process such as differential RNA stability. Recent promoter fusion studies, however, provide preliminary evidence that selective expression probably results from selective transcription (see below).

The pattern of *Antp* expression is shown as an example of a homeotic selector gene (*Figure 4*). As mentioned earlier, the *Antp* gene is required for the morphogenesis of structures derived from parasegment 4 (PS4) and the metathorax (or T3; see *Figure 11* for the relationship between segments and parasegments). Mutant embryos that completely lack *Antp* function show cuticle transformations whereby PS4 and PS5 acquire pattern elements normally associated with PS2. Mosaic analyses also indicate a localized requirement of *Antp* function within the thorax. Adult flies that contain patches of *Antp*⁻ tissue in the developing mesothorax show a transformation of legs into antennal structures, as well as malformations of the wings. *Antp*⁻ cells clones in head or abdominal regions do not cause discernible transformations or disruptions of these segments (77).

Antp transcripts are detected in a transverse band of nuclei within the presumptive thorax of cleavage cycle 14 embryos undergoing cellularization (*Figure 4a*). This site of expression corresponds to the presumptive PS4, which is the primary site of *Antp* function. By the beginning of gastrulation, about 30 min later, a second site of expression can be detected in a band of cells just posterior to the PS4 domain (*Figure 4b*). This second site of expression might not be parasegmental, but instead might correspond to the presumptive metathorax (T3). PS4 expression encompasses the entire circumference of the embryo, and includes both ectodermal and mesodermal tissues. In contrast, the T3 band is restricted to ventral regions, and ultimately becomes restricted to the mesoderm. During germ band elongation, there is a 'de-repression' or expansion of the *Antp* pattern (*Figure 4d* and *e*). *Antp* transcripts can be detected as anteriorly as PS3 and expression extends posteriorly through PS13. This expansion is a rather intriguing process since the initial *Antp* pattern corresponds quite closely with the genetic domain of *Antp* function. As indicated above, mosaic analyses do not reveal a

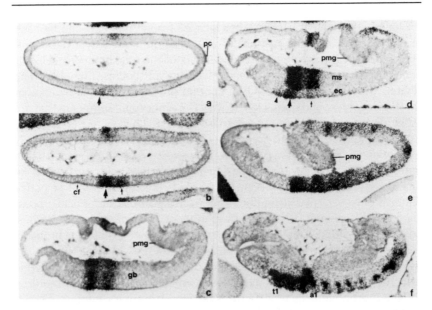

Figure 4. *Antp* expression in wild-type embryos. The embryos shown range in age from ~2 h (a) to ~9 h (f) after fertilization. All sections are sagittal and oriented with the anterior pole to the left and dorsal side up. (**a**) *Antp* expression in an embryo immediately prior to cellularization. A single band of expression is detected in the presumptive PS4 region (arrow). (**b**) Embryo undergoing gastrulation. *Antp* transcripts are detected in PS4 and T3 on the ventral surface (arrows) and in PS4 dorsally. (**c**) Embryo at the beginning of germ band elongation. Transcripts are detected in the same regions as in (b). (**d**) Embryo slightly older than that shown in (c). *Antp* transcripts are detected in the ectoderm of PS3 (arrowhead), all tissues of PS4 (large arrow) and in the mesoderm of T3. (**e**) Embryo near the completion of germ band elongation. *Antp* transcripts are detected throughout the germ band from PS3–PS14. (**f**) Embryo near the completion of germ band retraction. The highest levels of *Antp* transcripts are detected in the thorax. Lower levels extend posteriorly to the end of the germ band. a1, first abdominal segment; cf, cephalic furrow; ec, ectoderm; gb, germ band; ms, mesoderm; pc, pole cells; pmg, posterior midgut invagination; t1, first thoracic segment.

clear morphogenetic function for *Antp* products in posterior segments. During organogenesis, the primary, but not exclusive, site of *Antp* expression corresponds to the ventral cord of the central nervous system (CNS) (*Figure 4f*). Over the course of embryogenesis and larval development, there is a gradual restriction of *Antp* products to their initial limits of expression within the PS4 and T3 regions of the CNS (65).

The complexity of the *Antp* pattern shown in *Figure 4* is in part due to the occurrence of two promoters within the *Antp* transcription unit, called P1 and P2 (101,102). *Figure 4* shows the *Antp* pattern using a hybridization probe that detects transcripts derived from both promoters. The use of hybridization probes that detect RNAs made from each of the two promoters separately suggests that there might be tissue-specific differences in the activities of P1 and P2 (M.Scott, personal communication). Interestingly, cDNA sequence studies suggest that the proteins derived from both P1 and P2 transcripts possess an identical primary sequence (101,102) yet the P1 promoter maps nearly 70 kb upstream from the more proximal P2 promoter. The P1 and P2 RNAs primarily differ on the basis of a 1.7 kb compared to a 1.5 kb untranslated leader sequence.

During the first 1−2 h of expression, from cellularization through gastrulation, *Antp* RNAs are restricted to the nuclei of those cells that express the gene (65). Moreover, there is at least a 2 h lag between the time when *Antp* RNAs and proteins are detected during development (39,40). It is possible that the lag in appearance of cytoplasmic RNAs and protein synthesis reflects the very large size of the *Antp* transcription unit which is about 103 kb in length. It would take about 90 min for the synthesis of a primary transcript extending the length of the *Antp* unit. However, it is possible that the lag in appearance of *Antp* proteins results from a translational control mechanism especially since, as mentioned above, both the P1 and P2 transcripts contain unusually long untranslated leader sequences (101,102).

Due to the long lag in appearance of the *Antp* protein, the genetic circuitry controlling the initiation of *Antp* expression during early development is best examined by localizing RNAs using *in situ* hybridization, even though immunolocalization methods provide superior resolution and simpler three-dimensional imaging. Similar considerations pertain to the other homeotic selector genes, particularly *Ultrabithorax* (*Ubx*), which is about 75 kb in length (32). It is possible that even relatively 'small' homeotic genes are subject to translational control. For example, there is a lag of at least 2 h between the first appearance of *Sex combs reduced* (*Scr*) RNAs and the time when *Scr* proteins are first detected by immunolocalization (31). The *Scr* transcriptional unit is probably less than 20 kb in length.

3.2 Summary of homeotic expression patterns

Figure 5 summarizes the RNA distribution pattern for eight of the nine homeotic selector genes at the cellular blastoderm stage. As for *Antp*, there is a close correlation between embryonic sites of expression and domains of function based on genetic studies. Null mutations in each of these genes cause embryonic lethality and homeotic transformations. These initial limits of gene expression correspond quite closely to domains of gene function. Several of the genes show an expansion in expression during later stages, resulting in the appearance of homeotic products in tissues where they serve no obvious developmental function. For example, *Antp* products appear in abdominal segments of developing embryos, even though genetic studies indicate no clear role for *Antp* at these sites. In

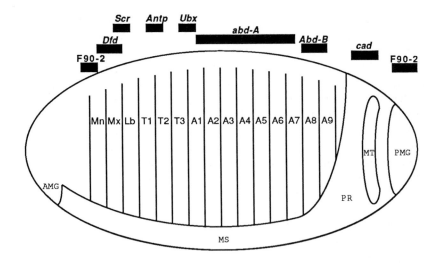

Figure 5. Primary domains of homeotic gene expression in an embryonic fate map. The fate map shows the origins of various embryonic tissues at the blastoderm stage of development. The solid bars above the map show the primary domains of expression for the eight homeotic lethal genes that contain a homeobox (*pb* is not included). The expression patterns shown correspond to those seen at the cellular blastoderm stage. The fate map is adapted from Hartenstein *et al.* (reviewed in ref. 13) and is oriented so that dorsal side is uppermost and the anterior pole is to the left. AMG, anterior midgut; MS, mesoderm; MT, malpighian tubules; PMG, posterior midgut; PR, proctodeum; Mn, mandibular head segment; Mx, maxillary segment; Lb, labium; T1 – T3, first through third thoracic segments; A1 – A9, first through ninth abdominal segments.

several cases, including *Antp*, there is an 'accordion' effect: after an expansion of expression there is a gradual sharpening during embryonic and larval development so that expression becomes restricted to more or less the original limits. The homeotic genes *Scr, Ubx* and *Abd-B* show an expansion of the original expression limits during the process of germ band elongation (103,104), and in the case of *Ubx* there is an accordion effect similar to that observed for *Antp* (105,106).

The homeotic selector gene *pb* (not shown in *Figure 5*) appears to be significantly different from the other eight genes (23). *pb⁻* embryos are fully viable and survive to adulthood. Moreover, the embryonic sites of *pb* expression do not appear to have a clear function. *pb* is required for the normal morphogenesis of several adult cuticular structures derived from the maxillary and labial head segments, and *pb⁻* mutants show transformations of mouth parts to legs. These structures are derived from the regions of the embryonic fate map where *pb* is expressed. However, *pb⁻* embryos do not show clear homeotic transformations or disruptions of these segments during embryogenesis., Such transformations appear much later in the life cycle and are evident only in adults (107).

A striking feature of *ANTP-C* and *BX-C* gene expression is the co-linearity between the physical order of the genes on the right arm of chromosome 3 and of the embryonic segments where they are expressed along the A/P body axis. Such co-linearity was first noted by Lewis for the *BX-C* genes (2). *Labial* (*lab*) corresponds to the proximal-most (closest to the centromere) gene of the *ANTP-C* (24 – 26) and is expressed in the anterior-most regions of the embryo; *Abd-B* is the distal-most gene of the *BX-C* and is expressed in the most posterior regions. This co-linearity does not play an obvious role in the expression of the homeotic selector genes; chromosomal rearrangements that disrupt the organization of the *BX-C* homeotic genes do not disrupt their normal activities (108). For example, *Abd-B* gene function is not detectably impaired when the *Ubx* gene is translocated onto a different chromosome. It has been proposed that an ancestral homeotic gene complex might have contained all of the *ANTP-C* and *BX-C* genes, as well as *cad*. Evidence for this possibility comes from recent genetic studies on the flour beetle, *Tribolius* (109). Separate mutations that disrupt the morphogenesis of very anterior and posterior body segments have been mapped to a single region of the *Tribolius* genome.

3.3 Unlinked regulatory genes maintain expression limits

Several genes have been shown to exert an indirect effect on segment identity by influencing the expression of the homeotic selector genes. *Polycomb* (*Pc*) was the first gene of this type, the homeotic regulatory class, to be identified and characterized (2,72). Extreme *Pc* mutants cause a strong transformation of all body segments towards the A8 phenotype, similar to that shown in *Figure 1c*. Double mutant studies showed that this homeotic transformation depends on $BX-C^+$ gene function (2,110). For example, $Pc^- - BX-C^-$ double mutants show a far less dramatic phenotype than that observed for Pc^- mutants. Furthermore, Pc^- embryos that contain extra doses of $BX-C^+$ genes show a more extreme transformation phenotype than do Pc^- embryos with the normal number of *BX-C* copies. This type of analysis prompted the proposal that *Pc* is required for restricting the activities of abdominal homeotic genes to their normal posterior domains of function. Pc^- embryos were proposed to express these posterior homeotic selector genes [*Ubx, abdominal-A* (*abd-A*) and *Abd-B*] in anterior segments, thereby causing the observed transformations. Direct support for this model has been obtained by localizing *BX-C* gene transcripts in Pc^- embryos (73). As predicted, each of the *BX-C* genes is expressed throughout the CNS of advanced-stage Pc^- embryos.

The mechanism by which Pc^+ products influence the expression of homeotic selector genes is not known. Lewis has proposed that *Pc* is a *trans*-acting factor which differentially represses the expression of the different *BX-C* genes. According to this model, there is a graded distribution of *Pc* products along the A/P axis with highest levels in anterior regions. Only *Ubx* can be expressed in anterior segments (such as the metathorax) since its promoter possesses a very low affinity for the *Pc* repressor. Other *BX-C* genes are not expressed at anterior sites since their promoters have a higher affinity for *Pc* products. Progressively more *BX-C* genes are expressed posteriorly due to lower levels of repressor. There is currently no evidence that supports or rejects this model.

An implication of the original Lewis model is that *Pc* products play a key role in the initiation of selective patterns of *BX-C* gene expression. This does not appear to be the case since the initial expression limits of each of the *BX-C* genes are normal in *Pc⁻* embryos (73; C.Wedeen and M.Levine, unpublished results). For example, during cellularization and gastrulation *Abd-B* RNAs are restricted to their normal sites of function in posterior body segments of *Pc* mutant embryos. Only in more advanced-stage embryos, during germ band elongation, is there ectopic expression of *Abd-B* in anterior segments. Interestingly, this misexpression occurs at about the time when a number of homeotic selector genes show expanded limits of expression during normal development. Thus, *Pc⁺* activity is required to maintain, but not establish, the correct limits of *BX-C* gene expression during development. Perhaps *Pc⁺* products serve to maintain the 'off state' of homeotic genes in cells where they are not initially expressed during early development.

Several other homeotic regulatory genes have been identified, including the maternally active gene *extra sex combs (esc)* (18,19). *esc⁻* embryos show a transformation phenotype quite similar to that observed for *Pc⁻* mutants (*Figure 1c*). Both its maternal inheritance and phenocritical period of activity suggested that *esc* might play a key role in the initiation, and not just the maintenance, of selective patterns of *ANTP-C* and *BX-C* expression. However, as seen for *Pc* mutants, the initial patterns of several *BX-C* genes, including *Ubx*, appear completely normal in early *esc⁻* embryos (111). Only later in development, during germ band elongation, is ectopic expression of *Ubx* seen in anterior segments. As for *Pc, esc⁺* products are required to maintain the off state of homeotic genes in cells where they are not initiated during early development. Only the gap and pair-rule classes of segmentation genes have been shown to be required for the initiation of the normal homeotic gene expression patterns (see below).

3.4 Posterior homeotic genes repress the expression of anterior genes

According to the Lewis model, the establishment of diverse pathways of segment morphogenesis involves different combinations of homeotic selector gene products (2). Posterior segment primordia are thought to contain progressively more homeotic gene products. To simplify the Lewis model, consider three hypothetical segments: A, B and C. Segment A contains homeotic gene product 1, segment B contains gene products 1 and 2, while segment C contains products 1, 2 and 3 (see *Figure 6a*). Morphological differences among the three segments are thought to be a direct consequence of these different permutations of homeotic gene expression. According to this model, homeotic transformations result from the removal of a particular homeotic gene product, thereby reiterating a specific combination of homeotic gene products in a segment primordium that usually contains a different combination. For example, a mutation that causes the loss of homeotic gene product 3 would result in the following permutation of products: A,1; B,1 and 2; C,1 and 2 (*Figure 6a*). Segments A and B contain their wild-type codes, but segment C now contains the combination of products characteristic of segment B. This would result in the transformation of segment C to segment B. Thus, the segments appear as A--B--B rather than A--B--C. The loss of product 3 from segment C 'unmasks' the activities of products 1 and 2 to give a segment B phenotype.

Recent molecular studies suggest that this model is not entirely accurate. Instead, it appears that the identity of a particular segment might involve high levels of a single homeotic gene rather than a combination of genes. Cross-regulatory interactions among homeotic selector genes play a role in restricting peak levels of expression to specific regions. The absence of a given homeotic selector gene product can cause the mis-expression of other homeotic genes. In general, it appears that homeotic genes expressed in posterior regions of the embryo repress the expression of genes located in more anterior regions. According to this view of cross-regulatory interactions, homeotic gene product 3 normally represses the expression of genes 1 and 2, and 2 represses 1. The loss of product 3 results in the posterior expansion of product 2 into segment C, thereby transforming segment C to a segment B phenotype (*Figure 6b*).

Traditional genetic studies could not distinguish between the two models diagrammed in *Figure 6a* and *b*. This problem has been recently addressed through the use of cloned homeotic DNAs which have been used as probes to examine the patterns of *ANTP-C/BX-C* expression in various homeotic mutants. On this basis, it has been shown that *Antp* is negatively regulated (either directly or indirectly) by the more posteriorly expressed *BX-C* genes, including *Ubx, abd-A*, and *Abd-B* (103,112). *Antp* transcripts are normally detected with highest levels in the PS4 region of the embryonic CNS, and with progressively lower levels of expression in more posterior ganglia (*Figure 6c*). However, in *BX-C⁻* embryos, *Antp* transcripts are detected at uniformly intense levels in the more posterior ventral ganglia, suggesting that *BX-C* genes normally exert a negative effect on *Antp* expression (*Figure 6c*). The *Antp* expression pattern observed in embryos that lack individual *BX-C* genes suggest that each of the three *BX-C* homeotic selector genes might independently

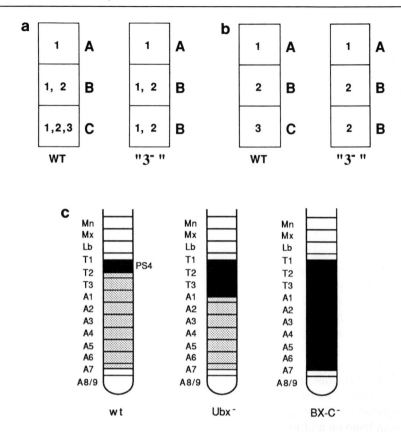

Figure 6. Posterior homeotic genes repress anterior genes. (**a**) Homeotic transformations according to the Lewis model. Consider three segments, A, B and C. Each of these segments is specified by a unique combination of homeotic gene products (designated 1, 2 and 3). Mutants lacking homeotic gene 3("3⁻") result in the transformation of segment C to a segment B phenotype due to the reiteration of the 'B code'. WT designates wild-type. (**b**) Homeotic transformations according to a cross-regulatory mechanism. Here, homeotic genes 1, 2 and 3 specify segments A, B and C, respectively. Mutants lacking gene 3 ("3⁻") result in the mis-expression of gene 2 in the segment C primordium, thereby transforming this segment to a B phenotype. The models shown in (a) and (b) are not mutually exclusive. (**c**) *Antp* expression in the CNS of wild-type (wt), *Ubx*⁻ and BX-C⁻ embryos. Solid areas indicate high levels of *Antp* expression, stippled areas indicate lower levels. In wt embryos, high levels of *Antp* expression are restricted to PS4, the posterior half of T1 and the anterior half of T2 (PS4), with lower levels extending posteriorly. In embryos that lack *Ubx* function, high levels of *Antp* expression extend posteriorly into the normal *Ubx* domain. In these embryos, *Antp* is strongly expressed in the posterior half of T1 through the anterior half of A1 (PS4−6), with lower levels extending posteriorly. In embryos that lack all known genes of the bithorax complex, high levels of *Antp* expression are seen from the posterior half of T1 through the anterior half of A7 (PS4−12).

repress *Antp*. For example, high levels of *Antp* transcripts are detected in the PS4 through PS6 regions of the CNS in *Ubx*⁻ (*abd-A*⁺, *Abd-B*⁺) mutants (*Figure 6c*). Similar studies show that *Ubx* is negatively regulated by *abd-A* and *Abd-B* (113). Regulatory interactions among *ANTP-C* and *BX-C* genes are not reciprocal. Although posterior genes can influence the expression of anterior genes, no case has been described where anterior genes repress posterior ones. For example, none of the *BX-C* genes show an anterior expansion in *Antp*⁻ embryos.

Regulatory interactions often involve the immmediate posterior neighbor of a given homeotic gene. For example, *Antp* is repressed by *Ubx*, which is in turn negatively regulated by *abd-A* (see *Figure 5*). However, there are several exceptions to this rule, such as the interactions between *Ubx* and the anterior homeotic gene *Scr*. *Scr* is normally expressed in PS2, while the primary site of *Ubx* expression corresponds to PS6. *Ubx* is also transiently expressed in the PS5 region of developing embryos, where it appears to play a role in repressing the expression of *Scr* (114). *Ubx*⁻ embryos show a homeotic transformation of PS5 into PS2. This interaction bypasses *Antp*, which is expressed in the region between the primary domains of *Scr* and *Ubx* expression. *Antp* exerts only a very weak effect on *Scr* expression. In addition, neither *Scr*⁻ nor *Antp*⁻ embryos show a posterior expansion of *Deformed* (*Dfd*) expression (43), and *abd-A* is not obviously influenced by *Abd-B* (73).

The abnormal patterns of *ANTP-C/BX-C* gene expression in *Pc*⁻ and *esc*⁻ mutants also provide support for the proposal that posteriorly expressed homeotic genes repress the expression of anterior genes. During germ band elongation, the *Antp*, *Ubx*, *abd-A* and *Abd-B* genes begin to show expanded patterns of expression. Within the next several hours these genes are expressed throughout the CNS of *Pc*⁻ or *esc*⁻ embryos (20,73,111). Over the course of subsequent development, *Antp* and *Ubx* expression gradually decline, whereas *abd-A* and *Abd-B* products persist at the original high steady-state levels. Mutant combination studies suggest that the repression of *Antp* and *Ubx* in advanced-stage *esc*⁻ and *Pc*⁻ embryos results from the mis-expression of the more posterior *BX-C* genes, *abd-A* and *Abd-B* (20,73). High levels of *Antp* products persist throughout the CNS of advanced-stage *Pc*⁻ embryos that lack all *BX-C* genes (*Pc*⁻ – *BX-C*⁻ double mutants). *Antp* products persist because of the removal of the *BX-C* genes (73). Similarly, *Ubx* products persist at high levels in *esc*⁻ – *abd-A*⁻ – *Abd-B*⁻ triple mutants (20).

3.5 The homeobox competition model

It is possible that cross-regulatory interactions among the *ANTP-C/BX-C* genes are mediated by the homeobox protein domains encoded by these genes (103). Homeobox proteins are likely to regulate gene expression at the level of transcription since they have been shown to bind specific

DNA sequences (115,116). Perhaps each of these genes autoregulates its own expression, thereby maintaining high levels of expression within its normal domain of function along the A/P embryonic body axis. A given homeotic gene might be restricted to a particular site within the embryo due to negative regulation by homeotic gene products in neighboring domains. For example, *Antp* proteins might bind to specific sites within the *Antp* promoter to augment the expression of this gene within the PS4/T3 region of the CNS. The proteins encoded by homeotic genes expressed in more posterior regions might 'compete' for the same binding sites within the *Antp* promoter that are recognized by the *Antp* protein and repress expression. As a result, the binding of the *Ubx*, *abd-A* and *Abd-B* proteins to the *Antp* promoter would interfere with maximal *Antp* expression in posterior regions of the CNS.

There are two critical predictions of this model:

(i)   different homeobox proteins recognize similar DNA sequences;
(ii)   *Antp* and *Ubx* autoregulate their own expression.

Direct proof that divergent homeobox proteins can recognize similar DNA sequences has been recently obtained (116), as shown in *Figure 7*. Each of four different full-length *Drosophila* homeobox proteins binds to the consensus sequence TCAATTAAAT. These proteins represent a broad spectrum of homeobox types in *Drosophila*, and no two share more than 50% amino acid identity within the homeobox.

Evidence for autoregulation is contradictory and fragmentary. A *Ubx* promoter fragment attached to the reporter gene *lacZ* shows only a limited response to endogenous *Ubx*[+] proteins when introduced into developing embryos by P-element-mediated germline transfer (93). In particular, this promoter drives high expression in the CNS of *Ubx*[-] embryos, suggesting that autoregulation is not a key aspect of the normal *Ubx* expression pattern. However, mesodermal expression in the PS7 region of the embryo disappears when this promoter is crossed into a *Ubx*[-] background. These data suggest that *Ubx* autoregulation might be important in some tissues, but possibly not in others. Autoregulation might also be unimportant for the maintenance of high levels of *Antp* expression in the PS4/T3 region of the CNS. Normal levels of *Antp* transcripts are detected in this region in *Antp*[-] embryos that do not detectably express the *Antp* protein (M.Scott and K.Harding, unpublished results).

4.   The *eve*, *ftz*, *en* regulatory circuit

4.1   *eve* and *ftz* define parasegmental boundaries

Regulatory interactions among three of the segmentation genes, *eve*, *ftz*, and *en* provide one of the best characterized models for the analysis of gene hierarchy during early development. *eve* and *ftz* are expressed in

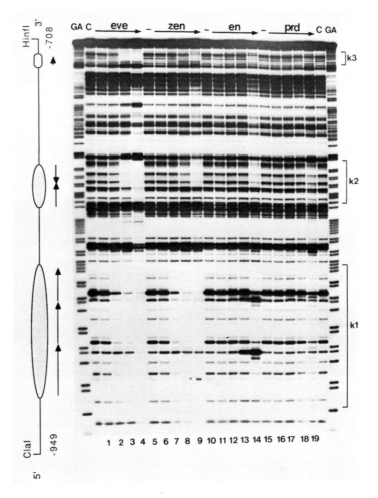

Figure 7. Comparative binding of four different homeobox proteins to 5' *en* sequences. A 240 bp fragment from a proximal region of the *en* promoter has been labeled with ^{32}P, and incubated with increasing amounts (five-fold increments) of *eve, zen, en* and *prd* protein extracts. Protein – DNA extracts have been treated with DNase I, and fractionated on a polyacrylamide/urea gel. There are three regions protected, designated k1, k2 and k3. The lanes labeled 'C' are controls that contain bacterial extracts lacking homeobox proteins. The lanes labeled 'GA' are Maxam – Gilbert sequencing reactions performed on the DNA fragment.

a series of seven periodic, transverse stripes along the A/P body axis of the early embryo (53,54,117,118). Each stripe initially includes 4 – 5 cells (at least the entire width of a parasegment primordium) and is separated from adjacent stripes by about 3 – 4 cells that show little or no expression. The stripes are transient and persist for a period of only 3 h from

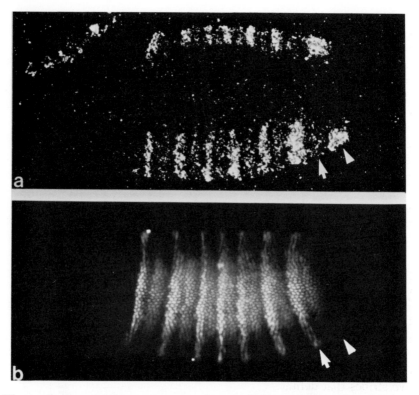

Figure 8. Complementary patterns of *eve* and *ftz* expression. Embryos are orientated with the anterior pole to the left and dorsal side uppermost. The embryo shown in (**a**) is a tissue section that was hybridized with a mixture of *eve* and *ftz* DNAs. The embryo shown in (**b**) displays the *eve* and *ftz* protein distribution patterns. The posterior-most *eve* stripe is indicated by an arrow, the posterior-most *ftz* stripe by an arrowhead. The 'zebra' expression patterns of *eve* and *ftz* are complementary at both the RNA and protein levels. Each *eve* stripe is anterior to the corresponding *ftz* stripe.

cellularization until the early periods of germ band elongation. By gastrulation, the *eve* and *ftz* stripes have 'sharpened' such that each includes only three cells and is separated from adjacent stripes by an interstripe region that includes five cells. *eve* and *ftz* are expressed in complementary cells, as shown in *Figure 8*. This complementarity of expression correlates quite well with previous genetic studies (26,119). Weak *eve* mutants cause the loss of the even-numbered body segments (actually, the odd-numbered parasegments), while *ftz*⁻ embryos lack the odd-numbered segments (even-numbered parasegments).

One of the primary functions of *eve* and *ftz* gene activity is to establish the 14-stripe *en* pattern during gastrulation (53,54,120,121) (*Figure 9a*). Recent DNA-binding studies are consistent with the notion that *eve* and *ftz* serve as transcription factors in the activation of *en* expression (115,

116). *en* is expressed in the posterior-most of the four cells that comprise each segment primordium and these sites define the posterior compartment and segment border (68,120,121). *eve* is required for the activation of the odd-numbered *en* stripes, while *ftz* is required for the activation of the even-numbered *en* stripes. These regulatory interactions might be direct since there is a close spatial and temporal linkage of the wild-type *eve, ftz* and *en* expression patterns. *en* stripes are detected within 10 min after the periodic *eve* and *ftz* patterns are established during cellularization. The anterior margin of each *eve* stripe coincides precisely with the odd-numbered *en* stripes (95,121). *ftz*⁻ embryos lack the even-numbered *en* stripes, and consequently, possess only one-half the normal number of segment borders at advanced stages (91). The *eve* pattern is not altered in *ftz*⁻ mutants, and the activation of the odd-numbered *en* stripes proceeds as in wild-type (53). As a result, advanced-stage *ftz*⁻ embryos lack pattern elements in alternating segments.

4.2 Why do *eve*⁻ embryos lack all 14 *en* stripes?

eve⁻ embryos lack all middle body segments, due to a general failure to activate any of the middle body *en* stripes during early development (53,54). This severe mutant phenotype is unique to *eve*; null mutations in the other seven pair-rule genes do not abolish segmentation but instead disrupt only every other segment (4–7,119). The loss of the odd-numbered *en* stripes is not surprising since there is an absence of the seven primary *eve* stripes that define the odd-numbered parasegments. What is more difficult to understand is the loss of the even-numbered *en* stripes. Their absence is puzzling since *ftz* expression appears nearly normal in *eve*⁻ embryos.

 eve⁺ products have been shown to accumulate at low levels within the *ftz* domain of wild-type embryos (54,118). It has been proposed that the failure to activate the even-numbered *en* stripes within *eve*⁻ embryos results from the loss of these weak, transient *eve* stripes (54). Perhaps the activation of the even-numbered *en* stripes requires the combined action of *ftz*⁺ products and low levels of *eve*. A major drawback to this proposal is the observation that *en* expression is initiated within the *ftz* domain prior to the detection of the weak *eve* stripes. Furthermore, while *eve*⁻ embryos fail to activate the even-numbered *en* stripes, these stripes reappear in *eve* mutants that also lack the pair-rule gene *odd-skipped*⁻ (*odd*) (120). Together, these results strongly suggest that the weak *eve* stripes are not critically required for the activation of *en*. A lack of function for the weak stripes would not be without precedent among pair-rule genes. Although the *paired* (*prd*) gene displays a 14-stripe pattern, only the original seven stripes appear to contribute directly towards segmentation (55). *prd*⁻ mutants show a typical pair-rule phenotype, suggesting that the second wave of *prd* stripes has no obvious function. Perhaps *prd*⁺ products, as well as *eve* and *ftz* products, can function only in certain regions of the embryo, and not in others. The additional *prd* and *eve* stripes might arise fortuitously.

The failure of *ftz* to activate the even-numbered *en* stripes in *eve* null mutants appears to result from a subtle disruption of the *ftz* pattern. Double staining of *eve* mutants with a mixture of *eve* and *ftz* antibodies shows that all seven *ftz* stripes are shifted anteriorly (95). Weak *eve* mutants cause relatively weak shifts, while strong mutants cause more severe shifts of the *ftz* pattern. This is particularly evident in a temperature-sensitive mutant of *eve*, as summarized in *Figure 9b* and *c*. At intermediate temperatures (25°C), the *eve* mutant causes a one-cell shift of the *ftz* pattern, resulting in partial pairwise alignments of adjacent

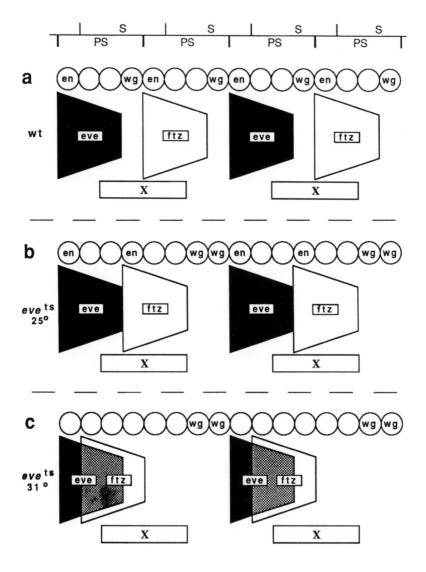

eve and *ftz* stripes. This abnormal spacing of *eve* and *ftz* causes a corresponding disruption of the *en* pattern. All 14 *en* stripes are activated, but there are pairwise alignments of adjacent stripes, such that stripes 1 + 2, and 3 + 4, etc. are brought closer together. The abnormal spacing of *en* stripes, as well as similar disruptions in other segment polarity genes, is probably responsible for the 'weak' *eve* phenotype, whereby the odd-numbered *en* stripes prematurely disappear during germ band elongation (95). *eve* and *ftz* are also required for the regulated expression of the segment polarity *wg* (69), which appears to encode a diffusable peptide with homology to growth factors (59,62). *wg* has been shown to influence the maintenance of the *en* pattern during advanced stages of development, presumably through a cell–cell signal transduction mechanism (see Section 4.5).

4.3 The regulation of *en* is a combinatorial process

There is much debate regarding the mechanism by which *eve* and *ftz* regulate *en* and other segment polarity genes (69,95,121). This regulatory process is complicated by the absence of a one-to-one correlation between the sites of *eve* or *ftz* expression and those cells that express *en*. As indicated in Section 4.1, each *eve* and *ftz* expression stripe initially encompasses four cells, while *en* is activated only within one of these cells. There are at least two opposing views, representing either a combinatorial or a threshold mechanism. According to a combinatorial model, *eve* and *ftz* act in concert with other factors, possibly other pair-rule gene products, to activate *en*. The cells that contain high levels of both *eve* (or *ftz*) and the other factor(s) correspond to the sites of *en* expression. Alternatively,

Figure 9. An anterior shift of the *ftz* zebra pattern in *eve* mutants results in altered *en* expression. The relationship between the patterns of *eve, ftz, en* and *wg* expression at gastrulation are shown in wild-type (wt) (**a**), an *eve*[ts] mutant at 25°C (**b**) and an *eve*[ts] mutant at 31°C (**c**). The bar labeled 'X' indicates the expression pattern of another pair-rule gene, possibly *odd-paired*. At the top of the figure, the relative positions of segments (S) and parasegments (PS) are indicated. (**a**) Wild-type: the *eve* and *ftz* patterns are complementary; each *eve* stripe is separated from the adjacent *ftz* stripes by one cell. *wg* is expressed in cells lacking either *eve* or *ftz* products. *en* is expressed in cells containing peak levels of *eve* or *ftz* products. The wild-type positioning of *en* results in the proper formation of evenly spaced segment boundaries. (**b**) *eve*[ts] at 25°C: the *ftz* expression pattern has shifted towards the anterior by one cell with respect to wild-type, whereas the expression of 'X' is unchanged. This results in an anterior shift of every other *en* stripe. In addition, half the *wg* stripes are eliminated and the remainder are broadened. The alteration in the *en* and *wg* expression patterns results in the pair-rule phenotype of these embryos. (**c**) *eve*[ts] at 31°C: the *ftz* pattern has shifted further to the anterior. The pattern of 'X' remains unchanged. The anterior *ftz* shift results in the elimination of half the *en* stripes. The remainder are also absent, due to malfunction of the mutant *eve* protein. The absence of *en* product results in the elimination of all segment boundaries and thus the continuous lawn of denticle hairs seen in these mutants.

it is possible that concentrations of *eve* or *ftz* above a certain threshold level are sufficient for activating *en*. By the time the *en* pattern is initiated, each *eve* and *ftz* stripe shows a graded distribution of product, with the anterior-most cell of each stripe showing peak levels (summarized in *Figure 9*). Perhaps *en* is activated only within the anterior-most cells, those expressing threshold levels.

The absence of all *en* stripes, including the *ftz*-domain *en* stripes, in strong *eve* mutants favors a combinatorial mechanism. As mentioned earlier, the *ftz* pattern is nearly normal in *eve*⁻ embryos, including the occurrence of threshold levels of *ftz* products (53,85). However, the *ftz* stripes are shifted to the anterior in *eve*⁻ embryos (95). Perhaps *ftz* is unable to function in the activation of *en* at these new locations. The simplest explanation for such regional specificity is the occurrence of a combinatorial mechanism for the regulation of *en*, as summarized in *Figure 9*.

According to this model, the periodic *ftz* pattern overlaps, but does not coincide, with another pair-rule gene, called X. *ftz* and X act in concert to activate *en*. Thus, only those cells that contain high levels of both products will come to express *en*. In weak *eve* mutants (i.e. the ts mutant at 25°C) the *ftz* pattern is shifted by only one or two cells, and still overlaps the domain of X expression. Consequently, *ftz* can function in these new positions to activate *en*. However, in null mutants, such as the ts mutant grown at 31°C, the *ftz* pattern is shifted outside the limits of X expression, and as a result there is a failure to activate the *ftz* domain *en* stripes. It has been proposed that the pair-rule gene *odd-paired* (*opa*) might act in concert with *ftz* to activate *en*, thereby fulfilling the role of gene X (69). Similarly, it is possible that *eve* acts in concert with a different pair-rule gene, possibly *paired*, to activate the odd-numbered *en* stripes (69).

4.4 Is an eve expression stripe a morphogen gradient?

The progressive shifting of the *ftz* pattern observed in an allelic series of *eve* mutants suggests that quantitative differences in the level of *eve* product within an expression stripe might be functionally significant. It has been shown that there is a correlation between the extent to which the *ftz* pattern is shifted in different *eve* mutants, and the level of *eve*⁺ product normally found in the affected cell (95). The weakest *eve* mutants cause a shifting of *ftz* into the posterior-most cells of the odd-numbered parasegments, which normally show the lowest level of *eve*⁺ products (see summary in *Figure 9*). More severe mutations cause a stronger anterior shift into regions that normally contain higher levels of product. Finally, null mutations disrupt *eve* activity in all the cells of an expression stripe, resulting in a severe anterior shift of the *ftz* pattern and the failure to activate the odd-numbered *en* stripes. These results show that there is a concentration-dependent response of *ftz* to *eve* products.

4.5 Segment polarity genes and cell – cell interactions

The establishment of the 14-stripe *en* pattern in the middle body region of the early embryo depends on regulatory interactions with transiently expressed pair-rule genes, such as *eve* and *ftz*. The maintenance of this pattern during advanced stages of development depends on a second tier of regulatory interactions, involving segment polarity genes (summarized in *Table 1*). It appears that these later interactions are independent of the transcriptional hierarchy governing the establishment of the *en* pattern during early development (63). At least one of the segment polarity genes, *wg*, does not encode a transcription factor, but instead appears to specify a signal involved in cell communication (59,62).

wg comes to be expressed in a series of 14 middle body stripes, with each site of expression just one cell anterior to each *en* stripe (69). Despite their location in different cells, *wg* exerts a profound regulatory effect on *en*. *wg* null mutations cause a premature loss of *en* expression following the completion of germ band elongation (63,64). Consistent with the notion of a two-tiered hierarchy, *wg* mutations have no discernible effect on the establishment of the *en* pattern at earlier stages. Because *wg* and *en* are not expressed in the same set of cells and, given the nature of the *wg* product, it would appear that *wg* influences *en* expression indirectly, through a cell communication mechanism. The segment polarity gene *patched* (*ptc*) exerts a very different effect on the *en* pattern. *ptc*⁻ embryos show ectopic sites of *en* expression, whereby 28 *en* stripes occur in the middle body region of advanced-stage mutants (63,64). The negative effect of *ptc* on *en* expression might be mediated by *wg* since *ptc – wg* double mutants show the same premature loss of *en* products as is observed for *wg* mutants. *Figure 10* summarizes the regulatory interactions among *ptc, wg*, and *en* (63). According to this model, *wg* specifies a diffusable factor that acts on neighboring cells to activate *en* expression. The action of *wg* is suppressed in those cells that express *ptc*. In this way a segmental polarity is maintained during development and *en* expression is restricted to the posterior compartment.

There is evidence that the late regulatory interactions among segment polarity genes are sufficient to initiate *en* expression independently of the pair-rule genes. A defective *en* promoter that is not expressed in the odd-numbered parasegments during early development comes to be expressed at these sites later, presumably in response to *wg* and other segment polarity genes (63). It has been proposed that the segment polarity regulatory program serves to give precise expression of late-acting segmentation genes such as *en* during advanced stages of development (63,64). Indeed, there is evidence that there are relatively frequent mistakes in the localization of the early *en* pattern. Over 95% of all cells that express *en* correspond to the anterior margins of *eve* and *ftz*, but 2 – 5% of the expressing cells do not (95,121). Perhaps later regulatory interactions with segment polarity genes are responsible for correctly positioning these mis-cued cells.

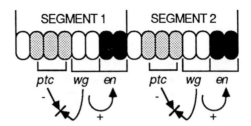

Figure 10. Interactions among *ptc, wg* and *en* during germ band elongation. Each segment is shown to be eight cells wide at this stage. Cells expressing *en* are shown in black, cells expressing *wg* in white and cells that have a requirement for *ptc* activity are stippled. Positive and negative interactions are indicated by + and − signs, respectively. The induction of *en* stripes during germ band elongation is independent of their earlier establishment during cellularization and gastrulation. This later induction requires *wg* expression in the cells immediately anterior to those that express *en. ptc* activity appears to be required to prevent *wg* from inducing *en* expression in cells in the anterior portion of each segment.

5.   Initiation of pair-rule stripes

5.1   Wild-type patterns of gap gene expression

Mounting evidence points to a key role for the gap genes in establishing striped patterns of segmentation gene expression during early development. Several lines of evidence suggest that the five gap genes (see *Table 1*) influence the segmentation pattern indirectly, through the regulation of pair-rule genes such as *ftz* and *eve* (84 – 86). It is possible that gap genes control development by directly modulating gene expression since at least two of these genes, *hb* and *Kr*, share homology with known transcription factors and contain multiple copies of a zinc finger DNA-binding motif (46,47).

The five known gap genes function in discrete, overlapping domains along the A/P embryonic body axis (4,92). The gap genes *hb, Kr* and *kni* act on progressively more posterior regions of the embryo, and together are responsible for establishing middle body segments, extending from the posterior region of the head through the eighth abdominal segment. The gap genes *tailless* (*tll*) and *giant* (*gt*) are required for the morphogenesis of certain head structures and are also required for the correct segmentation of posterior regions, including A8, A9 and the terminalia in the case of *tll* (122), and A4 through A7 in the case of *gt* (123). Gap mutants cause deletions of contiguous sets of segments in advanced-stage mutants and occasional mirror-image duplications of the remaining pattern elements. For example, the gap mutant *kni* causes the deletion of the first through seventh abdominal segments (4). *Kr⁻* embryos lack the first thoracic segment through the fifth abdominal

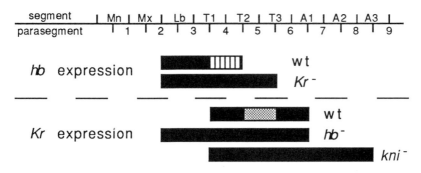

Figure 11. Cross-regulation among gap genes. The expression patterns of *hb* and *Kr* in wild type and gap mutant embryos during cell cycle 14 are shown. Only the middle body regions of the embryo are represented, as shown at the top of the figure. *hb* expression is indicated above the dashed line, and *Kr* expression below it. In wild-type (wt) embryos, *hb* expression is detected at high levels in the posterior half of PS2 and all of PS3 (solid bar); this band of expression extends posteriorly during the later stages of cell cycle 14 to include PS4 (cross-hatched bar). In the absence of *Kr* expression (Kr⁻), *hb* expression extends even further posteriorly to include PS5. In wild-type embryos, *Kr* expression is detected at the highest levels in PS4 and PS6 (solid bar); slightly lower levels of expression are detected in PS5 (stippled bar). In the absence of *hb* (hb⁻), *Kr* expression extends anteriorly to include the posterior half of PS2. In the absence of *kni*, *Kr* expression extends posteriorly through PS8. Mn, mandibular head segment; Mx, maxillary segment; Lb, labium.

segment, and there is a mirror-image duplication of the sixth abdominal segment. As indicated previously for *tll* and *gt*, gap genes do not necessarily act on a single contiguous region of the embryo. The gap gene *hb* is required for the establishment of both anterior and posterior body segments, including the labium and thorax as well as portions of the seventh and eight abdominal segments (124).

A summary of the domains of *hb* and *Kr* expression is presented in *Figure 11*. This summary includes only the thoracic and abdominal limits of expression and does not include the polar regions. These patterns are based on transcript localization studies using cloned *hb* and *Kr* DNAs as probes (67,92). The patterns of *hb* and *Kr* are highly dynamic and undergo rather dramatic changes over relatively short intervals of development. The summary in *Figure 11* is a 'snapshot' of embryos at the cellular blastoderm stage of development. An obvious feature of the wild-type expression patterns is that they do not coincide with cuticular defects seen in advanced-stage *hb⁻* and *Kr⁻* mutants. One reason for this discrepancy is the occurrence of cross-regulatory interactions among gap genes (92). A given gap mutant alters the expression of other gap genes. For example, *Kr* products are primarily restricted to the PS4–6 region of wild-type embryos. This site of expression corresponds to the region of the embryonic fate map that gives rise to the meso- and metathoracic segments (T1 and T2) and the first abdominal segment (A1). However, as indicated above, advanced-stage Kr^- embryos lack A2 – A5, as well as T2, T3 and A1. Thus, it appears that the domain of Kr gene function is twice as large as its wild-type site of expression in early embryos. As discussed below, this is due, at least in part, to altered patterns of hb and kni expression.

One of the central mysteries of the segmentation process is how these apparently simple paterns of expression give rise to an organized set of pair-rule stripes. At first glance it does not appear that these expression profiles possess sufficient informational content for the complex patterns of pair-rule gene expression. However, it should be emphasized that the gap patterns are dynamic and that the distribution of products is not uniform within a domain of expression. For example, the Kr protein shows a bell-shaped distribution pattern within the PS4 – 6 domain, with peak levels in PS5 (48). At a slightly later stage, this pattern becomes more complex such that peak levels are detected in PS4 and PS6, with a dip in expression at PS5 (88) (see *Figure 11*). Post-transcriptional regulation has been implicated in the generation of complex patterns of gap expression (48). For example, there are significant differences in the RNA versus protein profiles for Kr. Kr RNAs are detected not only in the PS4 – 6 region of the embryo, but also at the anterior and posterior poles (see *Figure 2a*). In contrast, Kr protein is not observed in polar regions, but is detected only in the PS4 – 6 region. It would appear that there is region-specific translation control of Kr RNAs, and that only those located in central regions of the embryo are translated and/or are stable and accumulate. There is some debate regarding the possibility that a similar post-transcriptional process is important for generating the bell-shaped distribution of Kr protein within the PS4 – 6 domain. Initial *in situ* hybridization experiments revealed a rather uniform distribution profile for Kr transcripts in this region (67). This apparently uniform pattern prompted the proposal that the Kr RNA is differentially translated (or that the protein is differentially stable) within the PS4 – 6 domain (48). However, recent *in situ* hybridization experiments reveal a more complex RNA pattern, which is similar to the protein profile that has been described (88). In particular, Kr RNAs show a transient bell-shaped pattern during cellularization. This would suggest that the establishment of an asymmetric Kr protein pattern is a problem of transcription or RNA stability.

The first systematic model that attempted to describe how continuous blocks of gap gene products give rise to periodic stripes was proposed by Meinhardt (89). The essential feature of the Meinhardt model is that borders, or regions of overlap, between adjacent domains of gap gene expression correspond to the sites where pair-rule stripes are initiated. A major drawback to this model is that there do not appear to be a sufficient number of gap borders to account for the periodic patterns of eight different pair-rule genes. However, this problem is simplified somewhat by recent genetic circuitry studies suggesting that the periodic patterns of only several of the pair-rule genes, the early class, are likely to be established in direct response to the gap genes (86,91). There are at least two ways in which relatively few gap genes can provide a lot of borders. Gap genes show highly dynamic expression patterns, raising the possibility that adjacent gap genes might form multiple, transient borders. For example, the first *Kr* transcripts that can be detected prior to cellularization are broadly distributed over middle body regions, and become restricted to the PS4 – 6 region later in development. Similarly, the location of the posterior margin of the anterior *hb* domain changes during early development, suggesting that there are at least two different *hb/Kr* boundaries during the course of normal development (125). Additional borders are also provided, at least in principle, by multiple domains of gap gene expression. As indicated above, *hb* is expressed in both anterior and posterior regions of the embryo, and *Kr* is expressed at the poles in addition to the PS4 – 6 domain.

5.2 Cross-regulatory interactions between gap genes

The maintenance of selective patterns of gap gene expression depends on cross-regulatory interactions whereby a given gap gene can influence the spatial limits of a neighboring gap gene (92). These regulatory interactions are reminiscent of the interactions that occur among homeotic genes. As discussed in Section 3.4, posterior homeotic genes such as *abd-A* and *Abd-B* repress the expression of the more anterior *Antp* and *Ubx* genes. In contrast to the posterior – anterior hierarchy observed for the homeotic genes, interactions among gap genes appear to be lateral. There is an anterior expansion of *Kr* expression in *hb⁻* embryos, and a posterior expansion of *hb* expression in *Kr⁻* embryos (summarized in *Figure 11*). Thus, it appears that *hb* exerts a negative effect on the expression of *Kr*, and *Kr* exerts a reciprocal effect on *hb*. Cross-regulatory interactions among homeotic genes are not reciprocal. For example, *Ubx* represses the expression of *Antp*, but the *Ubx* pattern is not altered in *Antp⁻* embryos.

Interactions among gap genes appear to be linear, and involve immediate neighbors, as is observed for the homeotic genes. For example, *Antp* expression expands posteriorly in *Ubx⁻* embryos, but is normal in *abd-A* mutants (see the summary of the domains of homeotic gene expression in *Figure 5*). Similarly, *hb* expression is disrupted by mutations in the neighboring *Kr* gene, but is not detectably changed in *kni* mutants. Since it is located in the center of the embryo, the *Kr* pattern is expanded by both *hb* and *kni* mutants (92). An additive expansion of the *Kr* limits is observed in *hb – kni* double mutants, such that *Kr* transcripts are distributed in both more anterior and posterior regions (88). It is likely that there is an anterior expansion of *kni* expression in *Kr* mutants and that *hb* mutants have no effect on the anterior margin of the *kni* pattern. However, there is currently only genetic evidence for this assertion since a *kni* probe has not yet become available for directly testing this possibility.

It is possible that there is a weak hierarchy governing regulatory interactions among the gap genes. There are temporal differences in the onset of localized patterns of gap gene expression, and it is possible that gap genes expressed early exert stronger regulatory effects than those expressed late. *hb* is unique among the gap genes in that it is expressed both maternally and zygotically (49). The other four gap genes are strictly zygotic, and those that have been examined are not expressed until cleavage cycle 10 – 11. In contrast, *hb* is expressed during oogenesis, and *hb* products are deposited in the unfertilized egg (124). Studies on the expression of pair-rule genes in gap mutants suggest that *hb* exerts a slightly earlier regulatory influence than the more posteriorly expressed gap genes such as *Kr* and *kni* (see Section 5.4). These studies also suggest that cross-regulatory interactions between gap genes is very sensitive to dosage, such that a two-fold reduction in the level of one gap gene product (a gap^+/gap^- heterozygote) can alter the expression of others.

5.3   Is autoregulation important?

Autofeedback mechanisms have been proposed to be important for the definition of discrete patterns of gene expression in response to more crudely localized positional cues. Such a mechanism has been invoked to account for the progressive localization of gap gene products to discrete domains along the embryonic body axis, and the definition of sharp boundaries between adjacent gap domains (89). As discussed above, the first *Kr* transcripts that are detected during early development are broadly distributed over the middle body region of the embryo (67). By the start of cell cycle 14, this band has narrowed and is restricted to the PS4 – 6 region. It seems quite plausible, at least in principle, that autoregulation would play a role in this refinement of the *Kr* pattern. Surprisingly, recent studies suggest that this is not the case, and that autoregulation is not important for the normal limits or levels of *Kr* expression (126).

As indicated earlier, nucleotide sequence analyses suggest that the *Kr* protein contains multiple copies of a zinc finger DNA-binding domain (46). Further support for this possibility was obtained by the recent characterization of a null allele of *Kr* (called Kr^9) (126). Advanced-stage Kr^9/Kr^9 homozygotes show the severe Kr^- phenotype, suggesting that these embryos completely lack active *Kr* product. The Kr^9 allele was shown to be a point mutation, resulting in a single amino acid substitution. This substitution replaces a cysteine residue with serine, and totally abolishes the activity of the *Kr* protein. Interestingly, this substitution resides within one of the putative zinc fingers, lending strong support to the notion that *Kr* functions as a sequence-specific DNA-binding protein. Anti-*Kr* antibodies were used to analyze expression in Kr^9

homozygotes. These studies showed that the mutant protein is stable and, more importantly, that its levels and limits of expression are virtually identical to that observed for the wild-type protein. This strongly suggests that autoregulation is not important for *Kr* expression. Of course, this does not rule out the possibility that other gap genes influence their own expression.

One can imagine a 'default' model for the establishment of the *Kr* pattern, whereby a ubiquitous factor(s) positively regulates *Kr*, and *hb* and *kni* repress its expression. Sharp *Kr* boundaries would depend only on sharp limits of *hb* and *kni* expression, which might be facilitated by the autoregulation of these latter genes. Similarly, it appears that autoregulation is not a general feature of all homeotic genes (see Section 3.5), although recent evidence suggests that at least one, *Dfd*, strongly influences its own expression (127). In any event, it does not appear that autoregulation is emerging as a key principle underlying segmentation gene expression.

5.4 Pair-rule expression in gap mutants

There are at least two very different mechanisms that can be envisaged for the establishment of a periodic series of pair-rule stripes. First, there might be an underlying periodicity in the early embryo due to maternal factors or early zygotic factors that span the length of the embryo. Past reaction/diffusion models have been proposed to account for the formation of periodic 'sine waves' along the length of the embryo, based on positional cues emanating from one or both embryonic poles (128). A second possibility is that the establishment of pair-rule stripes involves several locally restricted factors that act autonomously. As described above, the gap genes are likely candidates for such locally acting factors.

Initial studies on the wild-type patterns of *hairy* (*h*) (52), *ftz* (66) and *eve* (53,54) expression seemed to favor the global wave model. Each of the seven periodic stripes specified by a given pair-rule gene was found to appear at about the same time, during a very brief period of cleavage cycle 14 development. The stripes are quite regularly spaced and each encompasses about the same number of blastoderm cells. In fact, closer inspection of the onset of the *eve* and *ftz* stripes show that they do not arise at the same time, thereby raising the possibility that different stripes are subject to different regional control. For example, the *eve* protein is initially detected in all embryonic nuclei during cleavage cycle 12 (118). A localized staining pattern is first detected at the beginning of cleavage stage 14, whereby nuclei located in the posterior two-thirds of the embryo are more strongly stained than those located in the anterior third. A sharp anterior border appears at about 69% egg length (where 100% corresponds to the distance from the posterior pole to the anterior pole). Over the course of the next 10–15 min (in the absence of nuclear divisions) there is a gradual evolution of the seven-stripe *eve* pattern, and different stripes are formed at slightly different times. Stripe 1 is the first to appear, and shortly thereafter stripe 7 can be detected. At this time, two broad bands of *eve* expression can be observed in middle body regions, and these correspond to composite stripes 2 + 3, and 5 + 6. Stripes 2 and 3 separate, and then stripe 4 arises *de novo*. The last step in this process is the separation of stripes 5 and 6. This type of regional heterogeneity is also observed for the appearance of the different *h* and *ftz* stripes, although their exact order of appearance is somewhat different than that observed for *eve* (129).

Additional evidence that different pair-rule stripes are differentially regulated in a region-specific manner stems from studies on the patterns of pair-rule gene expression in different gap mutants. Mutations in any one of the five known gap genes cause region-specific disruptions of pair-rule gene expression (i.e. 85,86). *Figure 12a* shows the *eve* staining pattern in an early *kni⁻* mutant. Advanced-stage *kni* mutants lack the first through seventh abdominal segments. These segments are derived from the region of the embryonic fate map containing *eve* stripes 4, 5, and 6. Note that these stripes are absent in *kni* mutants, but that stripes 1, 2, 3, and 7 are normal. Other gap mutants disrupt different *eve* stripes. For example, *hb* mutants lack *eve* stripes 2 and 3. Similar region-specific disruptions of the *h* and *ftz* patterns have been reported for the gap mutants (84,85).

These results suggest that the regulation of different pair-rule stripes can be uncoupled and that the gap genes act on pair-rule genes in a region-specific manner. Furthermore, as noted above for *eve* in *kni* mutants, there is a close correlation between the altered patterns of pair-rule gene expression in a given gap mutant during early development and the segments that are deleted in the mutant during advanced stages of development. This provides strong support for a segmentation hierarchy, whereby gap genes control the segmentation pattern indirectly by regulating the expression of one or more pair-rule genes.

5.5 Early versus late pair-rule genes

There are a total of eight pair-rule genes in *Drosophila* (see *Table 1*). Of these, five have been directly shown to be expressed in a series of transverse stripes along the A/P embryonic body axis (52–56,66). It is likely that the other pair-rule genes will also prove to be expressed in periodic patterns. It is conceivable, but unlikely, that each of the eight pair-rule genes is directly regulated by gap gene products. A more reasonable alternative is that only a few 'early-acting' pair-rule genes directly respond to the gap genes, and that the periodic patterns of most pair-rule genes (i.e. 'late-acting') are established in response to the early genes. Genetic circuitry studies are compatible with the notion that the pair-rule genes can be divided into an early class and a late class (84–86) (see *Figure 3*). In order to understand the mechanism by which gap genes

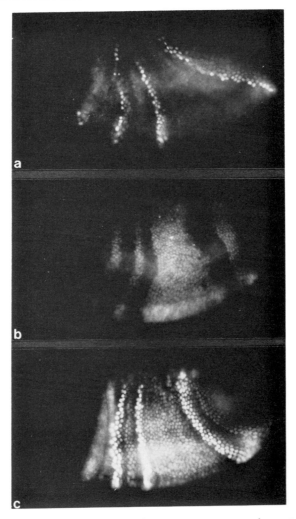

Figure 12. The *eve* and *ftz* expression patterns are complementary in *kni⁻* embryos. (**a**) *kni⁻* embryo stained with anti-*eve* antibody. Stripes 1, 2, 3 and 7 are normal, but 4, 5 and 6 are missing. (**b**) *kni⁻* embryo stained with anti-*ftz* antibody. Stripes 1, 2 and 7 are normal but 3–6 are fused into a broad, continuous band. (**c**) *kni⁻* embryo stained with a mixture of *eve* and *ftz* antibodies. The stronger bands correspond to *eve* and the weaker bands to *ftz*. It can be seen that the complementarity of *eve* and *ftz* expression is maintained in *kni⁻*.

specify stripes, it is important to identify likely target genes for their activities. The early pair-rule genes are currently the best candidates.

Cross-regulatory interactions among pair-rule genes have been shown to be important for their normal patterns of expression during early development (85,86). It is possible to assign pair-rule genes to early or late classes based on their cross-regulatory activities. An early gene would

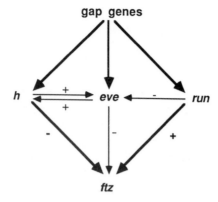

Figure 13. Interactions among the gap genes, *eve, runt (run), h* and *ftz*. In this diagram, interactions that are required for the establishment of expression patterns are indicated with thick arrows, interactions required for the refinement and maintenance of expression patterns are indicated with thin arrows. Positive and negative interactions are indicated by + and − signs, respectively. Gap gene activity is required for the establishment of the periodic expression patterns of the three early pair-rule genes, *h*, *eve* and *run*. These expression patterns appear to result from differential responses to different combinations of gap gene products. Once the early pair-rule patterns are established, they are maintained through cross-regulatory interactions: *h* is required to maintain *eve* expression at high levels and *run* prevents the *eve* stripes from extending into the even-numbered parasegments. Both *h* and *run* are required for the establishment of the wild-type *ftz* expression pattern. In the absence of *h*, the *ftz* stripes do not separate or sharpen properly, resulting in a series of 4–5 broad bands of *ftz* expression. In the absence of *run*, the *ftz* stripes are narrow and prematurely decay. *eve* function is required to prevent the *ftz* zebra pattern from shifting anteriorly but is not required for the establishment of the pattern.

be expected to strongly influence the establishment of a late pair-rule genes, but a reciprocal interaction should not be observed. By these criteria, the pair-rule genes *eve, h,* and *runt* belong to the early class, whereas *ftz* belongs to the late class (reviewed in ref. 130). *ftz* expression is strongly altered in *h* and *runt* mutants, and is also disrupted in *eve*⁻ embryos. In contrast, the *h* and *eve* patterns appear essentially normal in *ftz* mutants. These types of studies suggest that the initiation of the seven-stripe *ftz* pattern depends primarily on the pair-rule genes *h* and *runt*, while *eve* is important for the maintenance of the *ftz* pattern during later stages of development and for the correct positioning of the *ftz* stripes. These studies strongly suggest that the region-specific disruptions of the *ftz* pattern observed in each of the gap mutants are probably mediated by altered expression of *runt*, *h* and *eve*. The regulatory relationships between these genes is summarized in *Figure 13*.

Interactions between *h* and *ftz* provide one of the more thoroughly characterized examples of how one pair-rule gene can influence the expression of another (52,57,84,85,91,131). Several lines of evidence suggest that *h* represses *ftz* expression. In wild-type embryos the two genes are not expressed in the same set of cells, and in h^- mutants each of the *ftz* stripes broadens to include those cells that would normally contain h^+ products. The kinetics of the repression of *ftz* by *h* suggest that this interaction might be direct (131). h^+ products were mis-expressed in cells that normally express *ftz*. These experiments were performed by transforming flies with a construct in which the hsp70 promoter directs ectopic expression of the *h* protein. The mis-expression of *h* into the *ftz* domain resulted in the very rapid loss of *ftz* products. The mechanism by which h^+ products repress *ftz* is not known. It has been recently shown that the *h* protein accumulates in the nucleus, suggesting that it might regulate gene expression at the level of transcription (57). However, unlike other pair-rule genes that have been cloned and sequenced, *h* does not contain a homeobox (C.Rushlow, unpublished results) (see *Table 1*). Furthermore, the *h* protein coding sequence does not contain homology with any of the other well-characterized DNA binding motifs. Perhaps the *h* protein controls gene expression by modulating the activities of sequence-specific transcription factors, similar to the mechanism proposed for the EIA protein of adenovirus (132).

5.6 Complementary patterns of *eve* and *ftz* expression

eve and *ftz* provide an interesting model for understanding how two pair-rule genes are differentially regulated and come to be expressed in mutually exclusive sets of embryonic cells. The two genes are expressed in a complementary set of periodic stripes. The seven *eve* stripes define the odd-numbered parasegments, while the *ftz* stripes define the even-numbered parasegments. Despite their complementarity, the overall patterns of *eve* and *ftz* expression are remarkably similar. The seven-stripe patterns of *eve* and *ftz* expression arise at about the same time, during a brief interval of cleavage cycle 14 development (53,54,66). Both genes are expressed in middle body regions, and do not include the poles. The pair-rule expression of the two genes is transient, and persists for a period of only about 3 h during early development, extending from cellularization through germ band elongation. During this time there is a gradual refinement of the two patterns, such that each of the *eve* and *ftz* stripes sharpens. Initially, each stripe is about four cells in width, but there is a progressive loss of expression first in posterior, then more anterior, cells of each stripe. By the end of germ band elongation, *eve* and *ftz* are expressed only in the anterior-most cell of each of the original stripes. Finally, both genes are turned-off about the same time, but come back on in completely novel patterns during later stages of neurogenesis (117,118).

Two models have been proposed for the regulation of complementary

Table 2. Role of the pair-rule genes in the regulation of *eve* and *ftz* expression

Gene	Alters establishment of		Required for maintenance of	
	eve	*ftz*	*eve*	*ftz*
h	no	yes	yes	yes
run	no	yes	yes	yes
eve	no	no	yes	yes
ftz	no	no	no	yes
prd	no	no	no	no
opa	no	no	no	no
slp[a]	no	no	no	no
odd	no	no	no	no

[a]*slp*, sloppy-paired.

eve and *ftz* patterns, both of which are probably incorrect. First, a mutual exclusion model was proposed, whereby *eve* represses the expression of *ftz*, and *ftz* exerts a reciprocal effect on *eve*. Such a mechanism would insure the complementarity of the two patterns. According to the second model, complementary patterns of *eve* and *ftz* expression involve their differential regulation by a common set of transcription factors (86). A combination of factors that activates *eve* represses the expression of *ftz*, and vice versa. As discussed below, genetic circuitry studies exclude the first model and severely limit the second model.

   Table 2 summarizes studies carried out on the regulation of the two genes, which clearly assign *eve* to the early class of pair-rule genes and *ftz* to the late class. The mutual exclusion model for *eve* and *ftz* expression does not appear to be valid since the *eve* pattern is normal in *ftz*⁻ embryos (53). The classification of *eve* as an early pair-rule gene, one that might be directly regulated by gap and gene products, is based on the finding that none of the eight pair-rule mutants disrupts the establishment of the seven-stripe *eve* pattern (86). In contrast, the establishment of the *ftz* pattern is severely altered in *h* and *runt* mutants (85,91). Thus, the differential regulation model for the establishment of complementary *eve* and *ftz* patterns appears to be incorrect since only *eve* is likely to respond to the gap genes. However, it is possible that differential regulation is involved later in development for the refinement of the two expression patterns. *eve, h* and *runt* are all important for the maintenance of *eve* and *ftz* expression during more advanced periods of development (85,86). Moreover, it has been shown that each of these genes exerts opposite effects on *eve* and *ftz* (86). For example, as discussed above, *h*⁺ products repress *ftz* expression. In contrast, there is a premature loss of *eve* expression in gastrulating *h*⁻ embryos, suggesting that *h*⁺ products exert a positive effect on *eve* expression.

5.7 Distinct organizations of the *eve* and *ftz* promoters

Genetic circuitry studies suggest that the initiation of the two genes is likely to involve a different set of transcription factors, whereas a common set of factors might be responsible for the maintenance of their expression during later periods of development. Recent studies on the *cis*-regulatory elements contained within the *eve* and *ftz* promoters are consistent with this view. In particular, the promoter studies suggest that the establishment of the *eve* pattern depends on gap gene products, whereas the establishment of *ftz* involves other pair-rule genes.

Although *in situ* localization studies clearly demonstrate that the *eve* and *ftz* transcripts are distributed in a periodic pattern, this method is limited to the detection of steady-state RNAs and provides no information about *de novo* transcription. It is possible, in principle, that the striping process does not involve transcription, but instead involves a post-transcriptional process. For example, perhaps all embryonic cells efficiently transcribe a given pair-rule gene, but its mRNA is subject to differential degradation thereby permitting stable accumulation of high steady-state levels in stripes. The *eve* and *ftz* promoters have been examined by attaching 5′ flanking sequences to the report gene *lacZ*, and introducing these promoter fusions into flies by P-element-mediated germ line transformation (94,133,134). The most obvious and important conclusion of these studies is that striped expression of *eve* and *ftz* is regulated at the level of transcription.

Both the *eve* and *ftz* genes autoregulate their own expression (95,133,134). It is not clear whether such autoregulation contributes to the establishment of seven sharp stripes of gene expression during early development. At the very least, autoregulation is required to achieve maximal levels of *eve* and *ftz* expression within periodic limits established by regulatory interactions with other genes. The entire *ftz* promoter appears to be located within 6 kb of the 5′ flanking sequence. However, shorter 5′ fragments lack essential *ftz* promoter elements. For example, 4 kb of 5′ flanking sequence fails to give expression in lateral and dorsal regions of the early embryo (133). In this case, the seven-stripe pattern is restricted to ventral regions of the embryo. The loss of dorsal expression with short *ftz* promoter fusions is due to an autoregulatory element located between -6 and -4 kb upstream from the *ftz* transcription start site (134). A proximal region of the *ftz* promoter, located within the first 600 bp of the *ftz* 5′ end, is required for the establishment of the basic seven-stripe pattern (133). Expression driven by this proximal sequence, called the zebra element, is very weak and transient in lateral and dorsal regions, and will persist at high levels only if the distal autoregulatory element is present. The autoregulatory element has been shown to be dependent on *ftz*$^+$ gene activity. A heterologous promoter containing the -6 to -4 kb region gives seven stripes of *ftz* expression in a wild-type embryo (133). However, in *ftz*$^-$ embryos, this heterologous promoter is inactive.

The autoregulatory element has been shown to have properties of an enhancer. It has been suggested, but not yet proven, that *ftz* proteins directly bind to the upstream enhancer to drive optimal *ftz* expression.

Evidence for *eve* autoregulation was first provided by examining *eve* expression in a variety of *eve* mutants (95). Each of four different *eve* mutations were found to alter the *eve* pattern. These disruptions of the *eve* pattern are not due to lesions within the *eve* promoter since each of the mutations maps within the *eve* protein-coding sequence. The two null mutations that were examined result in a premature loss of *eve* expression in lateral regions of the embryo, similar to the alteration of the *ftz* pattern observed with *ftz* fusion promoters that lack the upstream enhancer element. *eve* promoter fusion studies have recently revealed an upstream autoregulatory element, located between about -6 and -5 kb upstream from the *eve* transcription start site (134).

5.8 The *eve* promoter contains separate potential gap response elements

Analyses of different *eve* promoter fusions reveal that the regulation of the different *eve* expression stripes can be uncoupled. A 5′ fragment that includes the first 5 kb of the *eve* promoter directs a highly abnormal pattern that includes only stripes 2, 3, and 7. Stripes 1, 4, 5, and 6 are absent (134). This observation suggests that the *cis*-regulatory elements needed for these latter stripes are located in a more distal region of the *eve* promoter. Additional *eve* promoter truncations uncouple the regulation of stripes 2, 3, 7. A 3 kb promoter expresses stripes 2 and 7, but lacks stripe 3. These studies have identified *cis*-regulatory elements within the *eve* promoter that are required for specific stripes (summarized in *Figure 14*). For example, the establishment of stripe 3 depends on sequences located between -5 and -3 kb upstream of the *eve* transcription start site.

An implication of these studies on the *eve* promoter is that the periodic *eve* pattern does not depend on a global cue spanning the length of the embryo, but instead involves several locally restricted factors that act autonomously. A similar conclusion was provided by studies on the *trans*-regulation of *eve* and other pair-rule genes (85,86). The gap genes are likely candidates for the region-specific regulation of different *eve* stripes, and it is possible that one or more gap gene products directly interact with specific sites within the *eve* promoter. For example, past studies of the genetic interactions suggest that *hb* gene activity is required for the establishment of stripes 2 and 3 (86), suggesting that hb^+ products might interact with one or more sites between -5 and -0.4 kb upstream from the *eve* start site.

The identification of separate *cis*-regulatory elements for different *eve* stripes is consistent with its classification as an early pair-rule gene based on the genetic interactions. *h* is the only other early pair-rule gene for

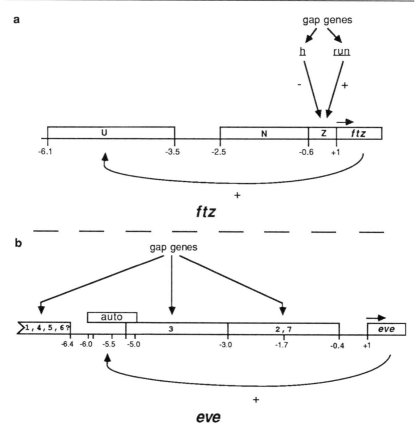

Figure 14. The *eve* and *ftz* promoters have a different organization. (**a**) The organization of the *ftz* promoter. Three *cis*-regulatory elements are required for the wild-type *ftz* expression pattern. The zebra element (Z), located between +1 and −600 bp, is necessary for the establishment of the seven *ftz* stripes. The upstream enhancer element (U), located between −3.5 and −6.1 kb, is required for maintaining high levels of *ftz* expression during advanced stages. The neurogenic element (N), located between −0.6 and −2.5 kb, is required for *ftz* expression in the nervous system. It is possible that the seven *ftz* zebra stripes are established through interactions of *h* and *run* products on the zebra element. *ftz* products have been shown to be required for the maintenance of its own expression by interacting, directly or indirectly, with the upstream element. (**b**) The organization of the *eve* promoter. This is different from that of *ftz*. No single *cis*-regulatory element has been found that is responsible for the establishment of all seven *eve* stripes. Instead, different sets of *eve* stripes require separate *cis* elements for their establishment. Sequences between −0.4 and −3.0 kb are required for the establishment of stripes 2 and 7. Sequences between −3.0 and −5.2 kb are required for the establishment of stripe 3. The *cis*-regulatory sequences required for the establishment of stripes 1, 4, 5 and 6 have not yet been identified. They could be located in more distal regions of the promoter, as shown, or at the 3′ end of the gene. It is possible that these *cis* elements respond to different combinations of gap gene products. Like *ftz*, the maintenance of the *eve* pattern depends on autoregulatory sequences in distal regions of the promoter ('auto'). Note that the *eve* and *ftz* coding regions are not drawn to scale.

which there is information regarding *cis*-regulation. The *h* promoter has been dissected genetically, whereby a number of different breakpoints have been mapped to different sites upstream of the *h* transcription start site (135). This study suggests that over 20 kb of 5' flanking sequence is required for the normal *h* pattern. Mutant breakpoints that map downstream of -20 kb can uncouple the expression of different *h* stripes. For example, a breakpoint located about 10 kb upstream gives a normal *h* pattern except that stripes 3 and 4 fail to form. A more proximal breakpoint, at about 8 kb upstream, causes the loss of stripes 6 and 7. The lack of stripes 3 and 4 observed for the -10 kb breakpoint is similar to the loss of these stripes in Kr^- embryos, suggesting that Kr^+ products might directly interact with distal sites of the *h* promoter.

The identification of separate *cis*-regulatory elements for different *eve* and *h* expression stripes distinguishes these promoters from that of *ftz*. *ftz* is the only late pair-rule gene promoter that has been examined. The establishment of the seven-stripe *ftz* pattern depends on only 600 bp of 5' flanking sequence. In no case has a truncated *ftz* promoter been shown to uncouple the regulation of different *ftz* stripes. This simple structure of the *ftz* promoter might reflect its relatively straightforward response to periodically distributed *hairy* and *runt* products (summarized in *Figure 13*). It is unlikely that gap gene products directly regulate *ftz* expression. The greater complexity of the *eve* and *h* promoters is consistent with the notion that these genes must 'decode' localized patterns of gap gene expression, and provide a greater level of positional specificity.

5.9 How do gaps make stripes?

The studies cited above provide important clues concerning the way in which gap gene products specify an organized set of pair-rule stripes along the embryonic body axis. Unfortunately, a solution to this complex problem appears to be a long way off. Significantly, both the *trans*-acting factors (gap proteins) and likely target promoters (early class pair-rule genes) have been identified.

It has been proposed that gap products control pair-rule promoters by a selective repression mechanism. This proposal is based on the injection of early embryos with drugs that block protein synthesis, and then examining pair-rule gene expression by *in situ* hybridization (136). If such drugs are injected just prior to the evolution of the seven-stripe *ftz* (or *eve*) pattern, a continuous band of expression spanning the length of the embryo is observed. This type of analysis suggests that pair-rule genes are activated by one or more global factors, and that pair-rule stripes evolve as a result of localized repression. Studies on the *eve* and *h* promoters are not entirely compatible with this model. A prediction of a strict repression model is that a truncated *eve* or *hairy* promoter fragment would give a broad band of expression, which would be later resolved into narrower stripes. This was observed in only one instance,

a

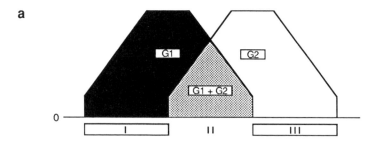

b

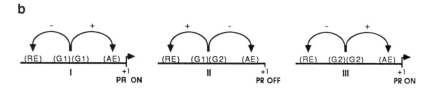

Figure 15. A model for the establishment of pair-rule stripes through differential responses to different combinations of gap gene products. (**a**) The distribution of two gap gene products (G1 and G2). Black areas indicate the presence of G1 products, white areas indicate the presence of G2 products and shaded areas indicate a region of overlap between G1 and G2. The distribution of G1 and G2 products results in three regions with different combinations of gap products (I, II and III). (**b**) The responses of a pair-rule gene promoter to these three combinations of gap products. The gap products might bind to *cis* sequences, as shown. When only G1 or G2 is present, a repressor element (RE) within the promoter is inhibited and an activator element (AE) is activated. This results in expression of the pair-rule gene in regions I and III. When both G1 and G2 are present, the repressor element is activated and the activator element is inhibited. This results in the repression of the pair-rule gene in region II. Similar on/off responses to other combinations of gap gene products along the A/P embryonic axis would result in a periodic pattern of pair-rule gene expression.

whereby a truncated *eve* promoter was initially expressed in a broad composite 2 + 3 band, and later resolved into stripes 2 and 3 (134). Overall, the *eve* and *hairy* promoter studies strongly suggest that the establishment of different *eve* and *h* stripes involves local activation by gap gene products (or combinations of products). A drawback of the drug injection studies is that the normally labile pair-rule RNAs are stabilized, possibly obscuring region-specific activation events.

Figure 15 summarizes a model for the generation of an organized set of pair-rule stripes and interstripes by different combinations of gap gene products. The essential feature of this model is that the interaction of gap gene products with an early pair-rule promoter controls gene expression by modulating the binding of both general activators and repressors. The diagram shows the expression of two different gap genes

(G1 and G2), which specify two pair-rule stripes (I and III) and one interstripe (II). According to this model, the two gap products have overlapping, bell-shaped distribution profiles. Perhaps the binding of G1 or G2 to the promoter stimulates the binding of general activators to the basal promoter (AE), and inhibits the binding of general repressors to a negative control element (RE). The dual regulation of the AE and RE elements might provide the basis for an amplified response of the promoter to crudely localized gap products. The binding of both G1 and G2 products to the promoter might have the opposite effect: the AE element is repressed and the RE element is stimulated, thereby giving an off state.

6. The initiation of homeotic stripes

6.1 Pair-rule genes influence the early homeotic patterns

The maintenance of selective patterns of homeotic gene expression within the CNS involves cross-regulatory interactions. Transcripts encoded by each *ANTP-C/BX-C* homeotic gene are first detected during cleavage stage 14, and by cellularization, they are restricted to well-defined spatial limits (reviewed in ref. 100). For the most part, these early localized expression patterns correlate quite closely to the genetic domains of *ANTP-C* and *BX-C* function. Several lines of evidence suggest that cross-regulatory interactions do not play a role in the establishment of these selective patterns of gene expression during early development. In general, mutations in a given homeotic gene do not alter the early expression patterns of other homeotic genes. For example, the initial *Antp* pattern is normal in *BX-C⁻* embryos, even though there is a dramatic breakdown in its normal limits of expression during later periods of development (103) (see summary in *Figure 6*). Similar studies indicate that unlinked regulatory loci such as *Pc* and *esc* are also required for the maintenance, but not establishment, of the *ANTP-C* and *BX-C* expression patterns (73,111).

The establishment of localized patterns of homeotic gene expression involves both gap genes and pair-rule genes. The initial patterns of *Scr*, *Antp* and *Ubx* expression have been shown to be altered in *ftz⁻* embryos (90). *ftz⁺* products appear to augment the expression of these homeotic genes, since they are expressed at much lower levels in early *ftz⁻* embryos as compared with wild-type embryos. Regulatory interactions among these genes might be relatively direct since their wild-type patterns of expression are tightly coupled. The early primary limits of *Scr*, *Antp* and *Ubx* correspond to parasegments 2, 4 and 6, respectively. These sites of expression coincide with *ftz* expression stripes 1, 2 and 3. *ftz* appears to exert a negative effect on the homeotic gene *Dfd* (43). In wild-type embryos the posterior margin of *Dfd* expression resides at the anterior margin of *ftz* stripe 1. There is a posterior expansion of *Dfd* products in *ftz⁻* embryos. The pair-rule genes *paired, odd-paired* and *eve* have been shown to exert a positive effect on *Dfd* expression (43).

6.2 The role of gap genes

Obviously, pair-rule gene products cannot be sufficient for non-periodic patterns of homeotic gene expression. If the initiation of homeotic genes involved only pair-rule gene products, then they would also be expressed in a series of stripes along the embryonic body axis. However, each homeotic gene is expressed in a restricted region of the embryo, encompassing one or just a few contiguous segmental (or parasegmental) units. It would appear that the establishment of these restricted limits of homeotic gene expression involves region-specific factors, possibly encoded by the gap genes.

Gap genes have been shown to be required for the normal limits of *Scr, Antp, Ubx* and *Abd-B* expression (31,40,84,87,88). In the case of *Scr* and *Ubx*, expression was monitored using antibodies. There is a significant delay in the time when the protein can be detected as compared with the time of appearance of their RNAs. Consequently, the abnormal patterns of *Scr* and *Ubx* that were observed could be an indirect effect that is mediated by altered expression of other early-acting genes. In any event, these studies indicate that *Scr* and *Ubx* expression depend on hb^+ gene activity.

Antp and *Abd-B* expression was monitored by the *in situ* localization of their RNAs, thereby permitting an evaluation of their expression soon after the time when the gap genes normally function during development (88). Examples of this analysis are shown in *Figure 16*. *Kr⁻* embryos show only a single site of *Antp* expression during gastrulation (*Figure 16a*). Wild-type embryos at this stage show two sites of *Antp* expression, corresponding to PS4 and T3 (see *Figure 4b*). There is a loss of the T3 domain in *Kr⁻*, suggesting that *Kr* exerts a positive effect on *Antp* expression. *Kr* appears to have the opposite effect on *Abd-B* expression, in that *Kr⁻* embryos show a second site of *Abd-B* expression within the presumptive thorax, in addition to its normal site of expression in the PS13–14 region (see *Figure 16b*). Thus, *Kr* appears to differentially regulate *Antp* and *Abd-B* and exert opposite regulatory effects on their expression. These regulatory effects might be relatively direct since the primary domain of Kr^+ gene activity encompasses the PS4–6 region of the embryonic fate map, which is where both the *Antp* and *Abd-B* patterns are disrupted in *Kr⁻* embryos.

The gap gene *kni* is also required for the normal limits of *Antp* and *Abd-B* expression. In contrast to the situation found for *Kr*, *kni* exerts a similar (negative) regulatory effect on the two genes. Advanced-stage *kni⁻* embryos lack the first through seventh abdominal segments, which derive from the region of the embryonic fate map that lies between the

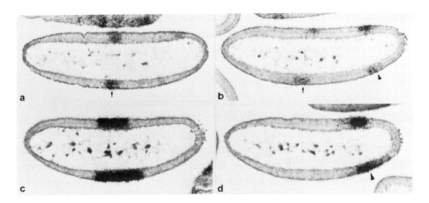

Figure 16. *Antp* and *Abd-B* expression in gap mutants. All embryos are 2.5 – 3 h old. Embryos shown in (**a**) and (**b**) are *Kr*⁻; embryos shown in (**c**) and (**d**) are *kni*⁻. (**a**) *Kr*⁻ embryo hybridized with an *Antp* RNA probe. Only the PS4 band of expression is seen (arrow); the T3 band of expression is absent (compare with *Figure 4b*). (**b**) *Kr*⁻ embryo hybridized with an *Abd-B* RNA probe. An ectopic band of expression is seen in the thorax (arrow), in addition to the normal posterior band (arrowhead). (**c**) *Antp* expression in *kni*⁻. A broad band of expression is detected in the thoracic region. This band is 2 – 3 times wider than that seen in a wild-type embryo at a similar stage. (**d**) *Abd-B* expression in *kni*⁻. *Abd-B* transcripts extend anteriorly from their wild-type domain (arrowhead). [Compare with the posterior band in (**b**)].

initial limits of *Antp* and *Abd-B* expression. *kni* mutants disrupt the early *Antp* and *Abd-B* patterns in this region of the embryo. Both patterns are broader than in wild-type, with the *Antp* pattern expanded posteriorly and the *Abd-B* pattern expanded anteriorly (see *Figure 16c* and *d*). Closer inspection of the altered *Antp* pattern reveals that the PS4 domain of expression is normal in *kni*⁻, whereas the T3 domain is expanded.

The occurrence of cross-regulatory interactions among gap genes (92) raises the possibility that one or more of these genes influence the *Antp* and/or *Abd-B* patterns only indirectly. For example, several lines of evidence suggest that *Kr*⁺ products exert a relatively direct effect in the activation of the *Antp* T3 domain of expression. It is possible that the expansion of the T3 limits seen in *kni* mutants is an indirect effect that is mediated by an altered pattern of *Kr* expression. It has been shown that there is a posterior expansion of the *Kr* pattern in *kni*⁻ mutants, prompting the proposal that *kni*⁺ products normally repress *Kr* and restrict its posterior boundary of expression. This expansion of the *Kr* pattern appears to cause a correlative expansion of the *Antp* T3 domain (88).

Diverse patterns of homeotic gene expression appear to involve their differential regulation by a common set of gap genes. For example, the *Antp* and *Abd-B* promoters might independently interpret the same gap gene products to give differential patterns of expression. The 'on state' of the *Antp* promoter depends, at least in part, on high concentrations of *Kr*⁺ products in the PS4 – 6 region of the early embryo. In contrast, these *Kr* products appear to be required for maintaining the *Abd-B* promoter in an 'off state' in this region of the embryo.

6.3 Gap genes versus pair-rule genes

The studies cited above suggest that a combination of gap gene products and pair-rule products define the parasegmental (or segmental) limits of homeotic gene expression during early development. At least two mechanisms can be envisaged: a lateral regulation model or a strict hierarchy model, as summarized in *Figure 17*. According to the lateral

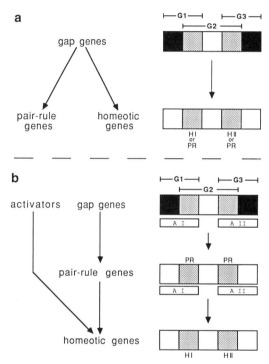

Figure 17. Two models for the establishment of pair-rule and homeotic gene expression patterns. In the lateral regulation model (**a**), the gap genes are directly responsible for establishing the limits of both pair-rule and homeotic gene expression. These limits are established through differential responses to different combinations of gap gene products. In the strict hierarchy model (**b**), the gap genes establish periodic patterns of pair-rule gene expression. The pair-rule genes then act, in concert with activators, to define the limits of homeotic gene expression. The activators could be either gap genes or maternal-effect segmentation genes. On the right-hand side of each diagram are shown, the distribution of gap gene (G1, G2 and G3), pair-rule gene (PR), homeotic gene (HI and HII) and the activator gene (AI and A11) products. Regions where the domains of gap gene expression overlap are shown in gray. Regions containing only one gap product are shown in black (G1, G3) or white (G2). The domains of activator gene expression are indicated below those of the gap and pair-rule genes.

regulation model (*Figure 17a*), different combinations of gap gene products directly establish parasegmental limits of both early pair-rule genes and homeotic genes. Periodic patterns of pair-rule gene expression involve a similar response of a given pair-rule promoter to different gap gene boundaries. The summary shown in *Figure 17* shows that pair-rule stripes 1 and 2 result from an identical response of the pair-rule gene to two different gap gene boundaries. The region of overlap between G1 and G2 is interpreted by this pair-rule promoter in the same way it responds to the G2 – G3 region. In contrast, the restricted expression of a homeotic gene to a single region of the embryo involves its differential interpretation of gap gene boundaries. Homeotic gene I (HI) is activated by the G1 – G2 boundary, but is either repressed or unaffected by the G2 – G3 boundary. Homeotic gene II (HII) shows the opposite response to these gap boundaries, thereby resulting in its restriction to a different region of the embryo. According to this model, sharp boundaries of homeotic gene expression emerge in direct response to the gap genes. Interactions with pair-rule gene products might serve to augment expression within these spatial limits. This model is quite similar to the way in which striped expression of *eve* and other early pair-rule genes might arise during early development (see *Figure 15*). The establishment of stripes involves a sophisticated 'decoding' of overlapping domains of gap gene products, and the stripes are maintained during later periods of development through autoregulation and interactions with other pair-rule genes.

The strict hierarchy model is summarized in *Figure 17b*. According to this model, interactions with pair-rule gene products do more than merely augment the levels of homeotic expression. Instead, sharp homeotic stripes are established through interactions with one or more pair-rule genes. Gap genes serve to define a relatively broad domain where a given pair-rule product is able to activate the target homeotic gene. For example, the limits of gap gene I (G1) expression define the activation domain for homeotic gene I (HI). A periodically distributed pair-rule gene product (PR) is able to activate HI only within the AI region. As a result, HI expression is restricted to a single region of the embryo, and is not periodically expressed.

6.4 A model homeotic promoter

Preliminary studies done on the regulation of *Scr* expression favor the strict hierarchy model discussed above, and might provide some general information on how a homeotic gene promoter is regulated by combinations of gap and pair-rule products. *Scr* products accumulate primarily in the PS2 region of early embryos, which corresponds to its primary domain of function (30,31,41) (see summary in *Figure 5*). During early embryogenesis, *Scr* is transiently expressed in a seven-stripe pattern, which coincides with the wild-type *ftz* pattern (90,104). However, by gastrulation, the first of these *Scr* stripes (corresponding to PS2) becomes

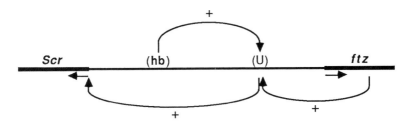

Figure 18. Model for the regulation of the homeotic gene *Scr* by a combination of pair-rule and gap gene products. In this model, the *ftz* upstream enhancer element (U) at a distance to regulate *Scr* transcription. The *Scr* and *ftz* coding regions are separated by ~20 kb, and are oriented in opposite directions, as shown. The diagram is not to scale. *Scr* is expressed in seven stripes during early stages of embryogenesis. These stripes are in register with the *ftz* zebra pattern. The anterior-most stripe (PS2) is more strongly expressed than the others and corresponds to the primary domain of *Scr* expression. It is possible that the *Scr* zebra pattern results from *cis*-regulation by the *ftz* upstream enhancer. In this model, *ftz* acts positively on the upstream element, which in turn activates *Scr* expression in a periodic manner. The higher levels of *Scr* expression seen in PS2 might result from a combined effect of *hb* and *ftz* on the upstream element. That is, in regions of the embryo that contain both *hb* and *ftz* products (i.e. PS2), the effect of the upstream element on *Scr* is further enhanced by the presence of *hb*.

more strongly expressed than the others, and by germ band elongation it is the only one to persist.

Studies of genetic interactions have identified two key factors in this regulated pattern of *Scr* expression, *hb* and *ftz* (31,90). *hb* mutations cause a sharp reduction in the levels of *Scr* expression. It is possible that expression would be completely abolished in true hb^- embryos, but this is difficult to evaluate since *hb* is expressed both during oogenesis and early embryogenesis. Only the zygotic component of *hb* expression has been evaluated in past studies of the genetic circuitry. *ftz* mutations also cause a reduction in *Scr* expression. A model for how *hb* and *ftz* products might interact to give strong PS2 expression of *Scr* is shown in *Figure 18*. According to this model, optimal *Scr* expression is achieved only when both *hb* and *ftz* products are bound to the *Scr* promoter.

Both *ftz* and *Scr* are located within the *ANTP-C*. In fact, the two genes are closely linked, and their 5′ ends map within 20 kb of each other (30). As discussed in Section 5.7, there is an upstream regulatory element (U) associated with the *ftz* promoter that mediates *ftz* autoregulation. It has been proposed that ftz^+ products interact with the U element to maintain maximum expression of the seven-stripe pattern during gastrulation and germ band elongation (133). Recent studies strongly suggest that *ftz* products act *in cis* on the *Scr* promoter to drive its transient seven-stripe pattern (137). That is, the interaction of *ftz* products with the U element might act over a considerable distance to direct transient, periodic expression of the neighboring *Scr* transcription unit. Perhaps because of the distance separating the U element and the *Scr* transcription start site, this expression is quite weak. The interaction of hb^+ products to a different region of the *Scr* promoter might enhance the effect of the U element on *Scr* expression. However, hb^+ products are expressed in the region from parasegments 0 through 3. The only region of overlap between the *hb* and *ftz* patterns corresponds to the first *ftz* stripe in PS2. Thus, strong *Scr* expression is restricted to this region. The other sites of *ftz* expression do not effect optimal expression of *Scr* due to the absence of *hb* products.

7. Concluding remarks

In summary, over 40 zygotically active genes control the differentiation of the early embryonic pattern. At least half of these genes encode proteins that contain the homeobox DNA-binding motif, and are likely to control cell fate by modulating gene expression at the level of transcription. The precise expression of these genes in specific subsets of embryonic cells involves a complex regulatory network, which has been defined, in part, by localizing their products in embryos carrying mutations for various regulatory genes. An important conclusion of these studies is that a given homeobox gene can influence the expression of others. Such cross-regulatory interactions have been implicated in the maintenance of selective patterns of homeotic gene expression during advanced stages of development, and play a role in the initiation of segment polarity genes in early embryos. Promoter fusion studies suggest that these interactions involve the regulation of *de novo* transcription. *In vitro* DNA-binding experiments are compatible with at least several of these regulatory interactions being direct; however, a major limitation of the genetic analyses is the uncertainty regarding the number of steps that intervene between the appearance of a homeobox protein (or combination of proteins) in a particular cell and the initiation (or repression) of its putative 'target' gene. A related drawback of the genetic analyses is the difficulty of identifying relatively ubiquitous factors that might act in concert with homeobox proteins to regulate gene expression. Clearly, a combination of genetic and biochemical assays will be required to resolve these issues.

The impact that homeobox proteins have on cell fate must involve the transcriptional control of target genes that directly specify cellular phenotype. The nature and identities of such targets are currently unknown, although the number of homeobox genes identified in *Drosophila* continues to expand. Nearly 50 homeobox genes have been identified to data, and it is conceivable that an additional 50 genes will be identified in the future. Given these numbers, it is possible that the homeobox 'superfamily' constitutes the predominant class of sequence-specific transcription factors in *Drosophila*. If so, it would not be surprising if as many as $10-25\%$ (or more) of the total number of genes in *Drosophila* are targets for homeobox proteins. Finally, the least understood, and one of the most intriguing, components of the segmentation hierarchy concerns the way in which the gap genes initiate striped expression of pair-rule genes and homeotic genes in the early embryo. Mounting evidence suggests that the gap genes are largely responsible for integrating the processes of segmentation and homeosis. At least two of the five known gap genes encode proteins that contain the zinc finger DNA-binding motif, suggesting that these proteins function as transcription factors. However, it is not clear whether the gap genes alone possess sufficient informational content to specify striped patterns of gene expression.

8. References

1. Van der Meer,J. (1977) Optical clean and permanent whole mount preparation for phase-contrast microscopy of cuticular structures of insect larvae. *Drosophila Inf. Serv.,* **52**, 160.
2. Lewis,E.B. (1978) A gene complex controlling segmentation in *Drosophila. Nature,* **276**, 565.
3. Kaufman,T.C., Lewis,R. and Wakimoto,B. (1980) Cytogenetic analysis of chromosome 3 in *Drosophila melanogaster*: the homeotic gene complex in polytene chromosome interval 84A-B. *Genetics,* **94**, 115.
4. Nusslein-Volhard,C. and Wieschaus,E. (1980) Mutations affecting segment number and polarity in *Drosophila. Nature,* **287**, 795.
5. Nusslein-Volhard,C., Wieschaus,E. and Kluding,H. (1984) Mutations affecting the larval cuticle in *Drosophila melanogaster* I: zygotic loci on the second chromosome. *Roux's Arch. Dev. Biol.,* **193**, 267.
6. Jurgens,G., Wieschaus,E., Nusslein-Volhard,C. and Kluding,H. (1984) Mutations affecting the larval cuticle in *Drosophila melanogaster* II: zygotic loci on the third chromosome. *Roux's Arch. Dev. Biol.,* **193**, 283.
7. Wieschaus,E., Nusslein-Volhard,C. and Jurgens,G. (1984) Mutations affecting the larval cuticle in *Drosophila melanogaster* III: zygotic loci on the X-chromosome and fourth chromosome. *Roux's Arch. Dev. Biol.,* **193**, 296.
8. Simpson,P. (1983) Maternal – zygotic gene interactions during formation of the dorsoventral pattern in *Drosophila* embryos. *Genetics,* **105**, 615.
9. Anderson,K. (1987) Dorsal – ventral embryonic pattern genes of *Drosophila. Trends Genet.,* **3**, 91.
10. Mohler,D. (1977) Developmental genetics of the *Drosophila* egg. I. Identification of 50 sex-linked cistrons with maternal effects on embryonic development. *Genetics,* **85**, 259.
11. Gans,M., Audit,C. and Masson,M. (1975) Isolation and characterization of sex-linked female sterile mutants in *Drosophila melanogaster. Genetics,* **81**, 683.
12. Schupbach,T. and Wieschaus,E. (1986) Maternal effect mutations altering the antero-posterior patterns of the *Drosophila* embryo. *Roux's Arch. Dev. Biol.,* **195**, 302.
13. Campos-Ortega,J.A. and Hartenstein,V. (1985) *The Embryonic Development of Drosophila Melanogaster.* Springer-Verlag, Berlin.
14. Levine,M., Harding,K., Wedeen,C., Doyle,H., Hoey,T. and Radomska,H. (1985) Expression of the homeobox gene family in *Drosophila*. Cold Spring Harbor Symp. Quant. Biol., **50**, 209.
15. Macdonald,P.M. and Struhl,G. (1986) A molecular gradient in early *Drosophila* embryos and its role in specifying the body pattern. *Nature,* **324**, 537.
16. Mlodzic,M. and Gehring,W.J. (1987) Expression of the *caudal* gene in the germ line of *Drosophila*; formation of an RNA and protein gradient during early embryogenesis. *Cell,* **48**, 465.
17. Ouweneel,W.H. (1976) Developmental genetics of homeosis. *Adv. Genet.,* **18**, 179.

18. Struhl,G. (1983) Role of the *esc*⁺ gene product in ensuring the selective expression of segment specific homeotic genes in *Drosophila. J. Embryol. Exp. Morphol.,* **76**, 297.

19. Struhl,G. and Brower,D. (1982) Early role of the *esc*⁺ gene product in the determination of segments in *Drosophila. Cell,* **31**, 285.

20. Struhl,G. and White,R.A.H. (1985) Regulation of the *Ultrabithorax* gene of *Drosophila* by other bithorax complex genes. *Cell,* **43**, 507.

21. Garcia-Bellido,A. (1975) Genetic control of wing disc development in *Drosophila*. In *Cell Patterning*. Elsevier/North-Holland, Amsterdam, pp. 161–182.

22. Sanchez-Herrero,E., Vernos,I., Marco,R. and Morata,G. (1985) Genetic organization of the *Drosophila* bithorax complex. *Nature,* **313**, 108.

23. Pultz,M.A., Diederich,R.J., Cribbs,D.L. and Kaufman,T.C. (1988) The *proboscipedia* locus of *Drosophila*: a molecular and genetic analysis. *Genes Dev.,* **2**, 901.

24. Hoey,T., Doyle,H.J., Harding,K., Wedeen,C. and Levine,M. (1986) Homeobox gene expression in anterior and posterior regions of the *Drosophila* embryo. *Proc. Natl. Acad. Sci. USA,* **83**, 4809.

25. Mlodzic,M., Fjose,A. and Gehring,W.J. (1988) Molecular structure and spatial expression of a homeobox gene from the *labial* region of the Antennapedia complex. *EMBO J.,* **7**, 2569.

26. Wakimoto,B.T., Turner,F.R. and Kaufman,T.C. (1984) Defects in embryogenesis in mutants associated with the Antennapedia gene complex of *Drosophila melanogaster. Dev. Biol.,* **102**, 147.

27. Regulski,M., Harding,K., Kostriken,R., Karch,F., Levine,M. and McGinnis,W. (1985) Homeobox genes of the Antennapedia and bithorax complex of *Drosophila. Cell,* **43**, 71.

28. Scott,M.P., Weiner,A.J., Polisky,B.A., Hazelrigg,T.I., Pirrotta,V., Scalenghe,F. and Kaufman,T.C. (1983) The molecular organization of the *Antennapedia* locus of *Drosophila. Cell,* **35**, 763.

29. Garber,R.L., Kuroiwa,A. and Gehring,W.J. (1983) Genomic and cDNA clones of the homeotic locus *Antennapedia* in *Drosophila. EMBO J.,* **2**, 2027.

30. Kuroiwa,A., Kloter,U., Baumgartner,P. and Gehring,W.J. (1985) Cloning of the homeotic *Sex combs reduced* gene in *Drosophila* and *in situ* localization of its transcripts. *EMBO J.,* **4**, 3757.

31. Riley,P.D., Carroll,S.B. and Scott,M.P. (1987) The expression and regulation of *Sex combs reduced* protein in *Drosophila* embryos. *Genes Dev.,* **1**, 716.

32. Bender,W., Akam,M., Karch,F., Beachy,P.A., Peifer,M., Spierer,P., Lewis,E.B. and Hogness,D.S. (1983) Molecular genetics of the bithorax complex in *Drosophila melanogaster. Science,* **221**, 23.

33. Karch,F., Weiffenbach,B., Peifer,M., Bender,W., Duncan,I., Celniker,S., Crosby,M. and Lewis,E.B. (1985) The abdominal region of the bithorax complex. *Cell,* **43**, 81.

34. McGinnis,W., Levine,M.S., Hafen,E., Kuroiwa,A. and Gehring,W.J. (1984) A conserved DNA sequence in homeotic genes of the *Drosophila* Antennapedia and bithorax complexes. *Nature,* **308**, 428.

35. Scott,M.P. and Weiner,A.J. (1984) Structural relationships among genes that control development: sequence homology between the *Antennapedia, Ultrabithorax* and *fushi tarazu* loci of *Drosophila. Proc. Natl. Acad. Sci. USA,* **81**, 4115.

36. White,R.A.H. and Wilcox,M. (1984) Protein products of the bithorax complex in *Drosophila. Cell,* **39**, 163.

37. Beachy,P.A., Helfand,S.L. and Hogness,D.S. (1985) Segmental distribution of bithorax complex proteins during *Drosophila* development. *Nature,* **313**, 545.

38. White,R.A.H. and Wilcox,M. (1985) Distribution of *Ultrabithorax* proteins in *Drosophila. EMBO J.,* **4**, 2035.

39. Wirz,J., Fessler,L.I. and Gehring,W.J. (1986) Localization of the *Antp* protein in *Drosophila* embryos and imaginal discs. *EMBO J.,* **5**, 3327.

40. Carroll,S.B., Laymon,R.A., McCutcheon,M.A., Riley,P.D. and Scott,M.P. (1986) The localization and regulation of *Antennapedia* protein expression in *Drosophila* embryos. *Cell,* **47**, 113.

41. Mahaffey,J. and Kaufman,T.C. (1987) Distribution of the *Sex combs reduced* gene products in *Drosophila melanogaster. Genetics,* **117**, 51.

42. Carroll,S.B., DiNardo,S., O'Farrell,P.H., White,R. and Scott,M.P. (1988) Temporal and spatial relationships between homeotic and segmentation gene expression in *Drosophila* embryos: distributions of the *fushi tarazu, engrailed, Sex combs reduced,*

Antennapedia and *Ultrabithorax* proteins. *Genes Dev.,* **2**, 350.

43. Jack,T., Regulski,M. and McGinnis,W. (1988) Pair-rule segmentation genes regulate the expression of the homeotic selector gene, *Deformed. Genes. Dev.,* **2**, 635.
44. Morata,G. and Lawrence,P.A. (1975) Control of compartment development by the *engrailed* gene in *Drosophila. Nature,* **255**, 614.
45. Kornberg,T. (1981) *engrailed*: a gene controlling compartment and segment formation in *Drosophila. Proc. Natl. Acad. Sci. USA,* **78**, 1095.
46. Rosenberg,U.B., Schroder,C., Preiss,A., Kienlin,A., Cote,S., Riede,I. and Jackle,H. (1986) Structural homology of the product of the *Drosophila Kruppel* gene with *Xenopus* transcriptional factor IIIA. *Nature,* **319**, 336.
47. Tautz,D., Lehmann,R., Schnurch,H., Schuch,R., Seifert,E., Kienlin,K. and Jackle,H. (1987) Finger protein of novel structure encoded by *hunchback*, a second member of the gap class of *Drosophila* segmentation genes. *Nature,* **327**, 383.
48. Gaul,U., Seifert,E., Schuh,R. and Jackle,H. (1987) Analysis of *Kruppel* protein distribution during early *Drosophila* development reveals posttranscriptional regulation. *Cell,* **50**, 639.
49. Gaul,U. and Jackle,H. (1987) How to fill a gap in the *Drosophila* embryo. *Trends Genet.,* **3**, 127.
50. Kuroiwa,A., Hafen,E. and Gehring,W.J. (1985) Cloning and transcriptional analysis of the segmentation gene *fushi tarazu* of *Drosophila. Cell,* **37**, 825.
51. Laughon,A. and Scott,M.P. (1984) Sequence of a *Drosophila* segmentation gene: protein structure homology with DNA-binding proteins. *Nature,* **310**, 25.
52. Ish-Horowicz,D., Howard,K.R., Pinchin,S.M. and Ingham,P.W. (1985) Molecular and genetic analysis of the *hairy* locus in *Drosophila. Cold Spring Harbor Symp. Quant. Biol.,* **50**, 135.
53. Harding,K., Rushlow,C., Doyle,H.J., Hoey,T. and Levine,M. (1986) Cross-regulatory interactions among pair-rule genes in *Drosophila. Science,* **233**, 953.
54. Macdonald,P.M., Ingham,P. and Struhl,G. (1986) Isolation, structure, and expression of *even-skipped*: a second pair-rule gene of *Drosophila* containing a homeobox. *Cell,* **47**, 721.
55. Kilcherr,F., Baumgartner,S., Bopp,D., Frei,E. and Noll,M. (1986) Isolation of the *paired* gene of *Drosophila* and its spatial expression during early embryogenesis. *Nature,* **321**, 493.
56. Gergen,J.P. and Butler,B.A. (1988) Isolation of the *Drosophila* segmentation gene *runt* and analysis of its expression during embryogenesis. *Genes Dev.,* **2**, 1179.
57. Carroll,S.B., Laughon,A. and Thalley,B.S. (1988) Expression, function and regulation of the *hairy* segmentation protein in the *Drosophila* embryo. *Genes Dev.,* **2**, 883.
58. Poole,S.J., Kauvar,L.M., Drees,B. and Kornberg,T. (1985) The *engrailed* locus of *Drosophila*: structural analysis of an embryonic transcript. *Cell,* **40**, 37.
59. Baker,N. (1987) Molecular cloning of sequences from *wingless*, a segment polarity gene in *Drosophila*: the spatial distribution of a transcript in embryos. *EMBO J.,* **6**, 1765.
60. Baumgartner,S., Bopp,D., Burri,M. and Noll,M. (1987) Structure of two genes at the *gooseberry* locus related to the *paired* gene and their spatial expression during *Drosophila* embryogenesis. *Genes Dev.,* **1**, 1247.
61. Cote,S., Preiss,A., Haller,J., Schuh,R., Kienlin,A., Seifert,E. and Jackle,H. (1987) The *gooseberry-zipper* region of *Drosophila*: five genes encode different spatially restricted transcripts in the embryo. *EMBO J.,* **6**, 2793.
62. Brown,A.M.C., Wildin,R.S., Prendergast,T.J. and Varmus,H.E. (1986) A retrovirus vector expressing the putative mammary oncogene *int*-1 causes partial transformation of a mammary epithelial cell line. *Cell,* **46**, 1001.
63. DiNardo,S., Sher,E., Heemskerk-Jorgens,J., Kassis,J.A. and O'Farrell,P.H. (1988) Two-tiered regulation of spatially patterned *engrailed* gene expression during *Drosophila* embryogenesis. *Nature,* **332**, 604.
64. Martinez-Arias,A. and White,R.A.H. (1988) *Ultrabithorax* and *engrailed* expression in *Drosophila* embryos mutant for segmentation genes of the pair-rule class. *Development,* **102**, 325.
65. Levine,M., Hafen,E., Garber,R.L. and Gehring,W.J. (1983) Spatial distribution of *Antennapedia* transcripts during *Drosophila* development. *EMBO J.,* **2**, 2037.
66. Hafen,E., Kuroiwa,A. and Gehring,W.J. (1984) Spatial distribution of transcripts from

the segmentation gene *fushi tarazu* of *Drosophila*. *Cell*, **37**, 825.

67. Knipple,D.C., Seifert,E., Rosenberg,U.B., Preiss,A. and Jackle,H. (1985) Spatial and temporal patterns of Kruppel gene expression in early *Drosophila* embryos. *Nature*, **317**, 40.

68. Kornberg,T., Siden,I., O'Farrell,P. and Simon,M. (1985) The *engrailed* locus of *Drosophila*: *in situ* localization of transcripts reveals compartment-specific expression. *Cell*, **40**, 45.

69. Ingham,P.W., Baker,N.E. and Martinez-Arias,A. (1988) Regulation of segment polarity genes in the *Drosophila* blastoderm by *fushi tarazu* and *even-skipped*. *Nature*, **331**, 73.

70. Martinez-Arias,A. and Lawrence,P. (1985) Parasegments and compartments in the *Drosophila* embryo. *Nature*, **313**, 639–642.

71. DiNardo,S., Kuner,J.M., Theis,J. and O'Farrell,P. (1985) Development of embryonic pattern in *D.melanogaster* as revealed by accumulation of the nuclear *engrailed* protein. *Cell*, **43**, 59.

72. Capdevila,M.P., Botas,J. and Garcia-Bellido,A. (1986) Genetic interactions between the *Polycomb* locus and the Antennapedia and bithorax complexes of *Drosophila*. *Roux's Arch. Dev. Biol.*, **190**, 339.

73. Wedeen,C., Harding,K. and Levine,M. (1986) Spatial regulation of Antennapedia and Bithorax gene expression by the *Polycomb* locus in *Drosophila*. *Cell*, **44**, 739.

74. Struhl,G. (1985) Near reciprocal genotypes caused by inactivation or indiscriminate expression of the *Drosophila* segmentation gene *ftz*. *Nature*, **318**, 677.

75. Schneuwly,S., Klemenz,R. and Gehring,W.J. (1987) Redesigning the body plan of *Drosophila* by ectopic expression of the homeotic gene *Antennapedia*. *Nature*, **325**, 816.

76. Wakimoto,B.T. and Kaufman,T.C. (1981) Analysis of larval segmentation in lethal genotypes associated with the Antennapedia gene complex of *Drosophila melanogaster*. *Dev. Biol.*, **81**, 51.

77. Struhl,G. (1981) A homeotic mutation transforming leg to antenna in *Drosophila*. *Nature*, **292**, 635.

78. Driever,W. and Nusslein-Volhard,C. (1988) A gradient of *bicoid* protein in *Drosophila* embryos. *Cell*, **54**, 89.

79. Driever,W. and Nusslein-Volhard,C. (1988) The *bicoid* protein determines position in the *Drosophila* embryo in a concentration-dependent manner. *Cell*, **54**, 95.

80. Klinger,M., Erdelyi,M., Szabad,J. and Nusslein-Volhard,C. (1988) Function of *torso* in determining the terminal anlagen of the *Drosophila* embryo. *Nature*, **335**, 275.

81. Frohnhofer,H.G. and Nusslein-Volhard,C. (1986) The organization of anterior pattern in the *Drosophila* embryo by the maternal gene *bicoid*. *Nature*, **324**, 120.

82. Tautz,D. (1988) Regulation of the *Drosophila* segmentation gene *hunchback* by two maternal morphogenetic centers. *Nature*, **332**, 281.

83. Nusslein-Volhard,C., Frohnhofer,H.G. and Lehmann,R. (1987) Determination of anterioposterior polarity in *Drosophila*. *Science*, **238**, 1675.

84. Howard,K. and Ingham,P. (1986) Regulatory interactions between the segmentation genes *fushi tarazu, hairy* and *engrailed* in the *Drosophila* blastoderm. *Cell*, **44**, 949.

85. Carroll,S.B. and Scott,M.P. (1986) Zygotically active genes that affect the spatial expression of the *fushi tarazu* segmentation gene during early *Drosophila* embryogenesis. *Cell*, **45**, 113.

86. Frasch,M. and Levine,M. (1987) Complementary patterns of *even-skipped* expression involve their differential regulation by a common set of segmentation genes in *Drosophila*. *Genes Devl.*, **1**, 981.

87. White,R.A.H. and Lehmann,R. (1986) A gap gene, *hunchback*, regulates the spatial expression of *Ultrabithorax*. *Cell*, **47**, 311.

88. Harding,K. and Levine,M. (1988) Gap genes define the limits of Antennapedia and Bithorax gene expression during early development in *Drosophila*. *EMBO J.*, **7**, 205.

89. Meinhardt,H. (1986) Hierarchical inductions of cell states: a model for segmentation in *Drosophila*. *J. Cell Sci. Suppl.*, **4**, 357.

90. Ingham,P.W. and Martinez-Arias,A. (1986) The correct activation of Antennapedia and bithorax complex genes requires the *fushi tarazu* gene. *Nature*, **324**, 592.

91. Ingham,P.W., Ish-Horowicz,D. and Howard,K.R. (1986) Correlative changes in homeotic and segmentation gene expression in *Kruppel* mutant embryos of *Drosophila*. *EMBO J.*, **5**, 1659.

92. Jackle,H., Tautz,D., Schuh,R., Seifert,E. and Lehmann,R. (1986) Cross-regulatory interactions among the gap genes of *Drosophila*. *Nature,* **324**, 668.
93. Bienz,M. and Tremml,G. (1988) Domain of *Ultrabithorax* expression in *Drosophila* visceral mesoderm from autoregulation and exclusion. *Nature,* **333**, 576.
94. Hiromi,Y. and Gehring,W.J. (1987) Regulation and function of the *Drosophila* segmentation gene *fushi tarazu*. *Cell,* **50**, 963.
95. Frasch,M., Warrior,R., Tugwood,J.D. and Levine,M. (1988) Molecular analysis of *even-skipped* mutants in *Drosophila* development. *Genes Dev.,* **2**, 1824.
96. Berleth,T., Burri,M., Thoma,G., Bopp,D., Richstein,S., Frigerio,G., Noll,M. and Nusslein-Volhard,C. (1988) The role of localization of *bcd* RNA in organizing the anterior pattern of the *Drosophila* embryo. *EMBO J.,* **7**, 1749.
97. Duncan,I. (1987) The bithorax complex. *Annu. Rev. Genet.,* **21**, 285.
98. Doyle,H.J., Harding,K., Hoey,T. and Levine,M. (1986) Transcripts encoded by a homeobox gene are restricted to dorsal tissues of *Drosophila* embryos. *Nature,* **323**, 76.
99. Rushlow,C., Doyle,H., Hoey,T. and Levine,M. (1987) Molecular characterization of the *zerknullt* region of the Antennapedia gene complex in *Drosophila*. *Genes Dev.,* **1**, 1268.
100. Akam,M. (1987) The molecular basis for metameric pattern in the *Drosophila* embryo, *Development,* **101**, 1.
101. Laughon,A., Boulet,A.M., Bermingham,J.R., Laymon,R.A. and Scott,M.P. (1986) The structure of transcripts from the homeotic *Antennapedia* gene of *Drosophila*: two promoters control the major protein-coding region. *Mol. Cell. Biol.,* **6**, 4678.
102. Stroeher,V., Jorgensen,E.M. and Garber,R.L. (1986) Multiple transcripts from the *Antennapedia* gene of *Drosophila melanogaster*. *Mol. Cell Biol.,* **6**, 4667.
103. Harding,K., Wedeen,C., McGinnis,W. and Levine,M. (1985) Spatially regulated expression of homeobox genes in *Drosophila*. *Science,* **229**, 1236.
104. Martinez-Arias,A., Ingham,P.W., Scott,M.P. and Akam,M.E. (1987) The spatial and temporal deployment of *Dfd* and *Scr* transcripts throughout development of *Drosophila*. *Development,* **100**, 673.
105. Akam,M. (1983) The location of *Ultrabithorax* transcripts in *Drosophila* tissue sections. *EMBO J.,* **2**, 2075.
106. Akam,M. and Martinez-Arias,A. (1985) The distribution of *Ultrabithorax* transcripts in *Drosophila* embryos. *EMBO J.,* **44**, 1689.
107. Kaufman,T.C. (1978) Cytogenetic analysis of chromosome 3 in *Drosophila melanogaster*. Isolation and characterization of four new alleles of the *proboscipedia* (*pb*) locus. *Genetics,* **90**, 579.
108. Struhl,G. (1984) Splitting the bithorax complex of *Drosophila*. *Nature,* **308**, 454.
109. Beeman,R.W. (1987) A homeotic gene cluster in the red flour beetle. *Nature,* **327**, 247.
110. Duncan,I.M. and Lewis,E.B. (1982) Genetic control of body segment differentiation in *Drosophila*. In *Developmental Order: its Origin and Regulation*. Subtelney,S. (ed.), Alan R.Liss, New York, pp. 533–544.
111. Struhl,G. and Akam,M. (1985) Altered distributions of *Ultrabithorax* transcripts in *extra sex combs* mutant embryos in *Drosophila*. *EMBO J.,* **4**, 3259.
112. Hafen,E., Levine,M. and Gehring,W.J. (1984) Regulation of *Antennapedia* transcript distribution by the bithorax complex in *Drosophila*. *Nature,* **307**, 287.
113. Struhl,G. and White,R.A.H. (1985) Regulation of the *Ultrabithorax* gene of *Drosophila* by other bithorax complex genes. *Cell,* **43**, 507.
114. Struhl,G. (1982) Genes controlling segmental specification in the *Drosophila* thorax. *Proc. Natl Acad. Sci. USA,* **79**, 7380.
115. Desplan,C., Theis,J. and O'Farrell,P. (1985) The *Drosophila* developmental gene, *engrailed*, encodes a sequence specific DNA binding activity. *Nature,* **318**, 630.
116. Hoey,T. and Levine,M. (1988) Divergent homeobox proteins recognize similar DNA sequences in *Drosophila*. *Nature,* **332**, 858.
117. Carroll,S.B. and Scott,M.P. (1985) Localization of the *fushi tarazu* protein during *Drosophila* embryogenesis. *Cell,* **43**, 47.
118. Frasch,M., Hoey,T., Rushlow,C., Doyle,H. and Levine,M. (1987) Characterization and localization of the *even-skipped* protein of *Drosophila*. *EMBO J.,* **6**, 749.
119. Nusslein-Volhard,C., Kluding,H. and Jurgens,G. (1985) Genes affecting the segmental subdivision of the *Drosophila* embryo. *Cold Spring Harbor Symp. Quant. Biol.,* **50**, 145.
120. DiNardo,S. and O'Farrell,P.H. (1987) Establishment and refinement of segmental

pattern in the *Drosophila* embryo: spatial control of *engrailed* expression by pair-rule genes. *Genes Dev.,* **1**, 1212.

121. Lawrence,P.A., Johnston,P., Macdonald,P. and Struhl,G. (1987) The *fushi tarazu* and *even-skipped* genes delimit the borders of parasegments in *Drosophila* embryos. *Nature,* **328**, 440.

122. Strecker,T.R., Kongsuwan,K., Lengyel,J.A. and Merriam,J.R. (1986) The zygotic mutant *tailless* affects the anterior and posterior ectodermal regions of the *Drosophila* embryo. *Dev. Biol.,* **113**, 64.

123. Petscheck,J.P., Perrimon,N. and Mahowald,A.P. (1987) Region specific defects in *1(1)giant* embryos of *Drosophila melanogaster*. *Dev. Biol.,* **119**, 175.

124. Lehmann,R. and Nusslein-Volhard,C. (1987) *hunchback*, a gene required for segmentation of an anterior and posterior region of the *Drosophila* embryo. *Dev. Biol.,* **119**, 402.

125. Schroder,C., Tautz,D., Seifert,E. and Jackle,H. (1988) Differential regulation of the two transcripts from the *Drosophila* gap segmentation gene *hunchback*. *EMBO J.,* **7**, 2881.

126. Redemann,N., Gaul,U. and Jackle,H. (1988) Disruption of a putative Cys–zinc interaction eliminates the biological activity of the *Kruppel* finger protein. *Nature,* **332**, 90.

127. Kuziora,M.A. and McGinnis,W. (1988) Autoregulation of a homeotic selector gene. *Cell,* **55**, 477.

128. Turing,A. (1952) The chemical basis of morphogenesis. *Phil. Trans. Roy. Soc. Lond. B,* **237**, 37.

129. Weir,M.P. and Kornberg,T. (1985) Patterns of *engrailed* and *fushi tarazu* expression reveal novel intermediate stages in *Drosophila* segmentation. *Nature,* **318**, 433.

130. Ingham,P.W. (1988) The molecular genetics of embryonic pattern formation in *Drosophila*. *Nature,* **335**, 25.

131. Ish-Horowicz,D. and Pinchin,S.M. (1987) Pattern abnormalities induced by ectopic expression of the *Drosophila* gene *hairy* are associated with repression of *ftz* transcription. *Cell,* **51**, 405.

132. Hoeffler,W.K., Kovelman,R. and Roeder,R.G. (1988) Activation of transcription factor IIIC by the adenovirus E1A protein. *Cell,* **53**, 907.

133. Hiromi,Y., Kuroiwa,A. and Gehring,W.J. (1985) Control elements of the *Drosophila* segmentation gene *fushi tarazu*. *Cell,* **43**, 603.

134. Harding,K., Hoey,T., Warrior,R. and Levine,M. (1989) Autoregulatory and gap gene response elements of the *even-skipped* promoter of *Drosophila*. *EMBO J.,* in press.

135. Howard,K., Ingham,P. and Rushlow,C. (1988) Region specific alleles of the *Drosophila* pair-rule gene *hairy*. *Genes Dev.,* **2**, 1037.

136. Edgar,B., Weir,M.P., Schubiger,G., and Kornberg,T. (1986) Repression and turnover pattern of *fushi tarazu* RNA in the early *Drosophila* embryo. *Cell,* **47**, 747.

137. Rushlow,C. and Levine,M. (1988) Combinatorial expression of a *ftz–zen* fusion promoter suggests the occurrence of *cis*-interactions between genes of the ANT-C. *EMBO J.,* **7**, 3479.

138. Fjose,A., McGinnis,W. and Gehring,W.J. (1985) Isolation of a homeobox-containing gene from the *engrailed* region of *Drosophila* and the spatial distribution of its transcripts. *Nature,* **313**, 284.

139. Regulski,M., McGinnis,W., Chadwick,R. and McGinnis,W. (1987) Developmental and molecular analysis of *Deformed*; a homeotic gene controlling *Drosophila* head development. *EMBO J.,* **6**, 767.

140. Levine,M. and Hoey,T. (1988) Homeobox proteins as sequence specific transcription factors. *Cell,* **55**, 537–540.

141. Rhodes,D. and Klug,A. (1988) Zinc fingers: a novel motif for nucleic acid binding. In *Nucleic Acids and Molecular Biology*. Eckstein,F. and Lilley,D.M.J., (eds), Springer-Verlag, Berlin, Volume 2.

3

Caenorhabditis
Kenneth J.Kemphues

1. Introduction

Classical embryological studies of nematodes, primarily by Van Beneden and Boveri near the turn of the century, have made lasting contributions to our understanding of embryonic development (1). However, during most of this century, nematodes have been eclipsed as a model system for embryology by organisms with more tractable embryos such as sea urchins, insects, amphibians, birds, and mice. Two features of the free-living soil nematode *Caenorhabditis elegans* have returned nematodes to a prominent place in embryological investigations: its suitability for genetic analysis and its invariant and completely described cell lineage. These two features, combined with technological advances in microscopy and molecular biology, are providing the opportunity to combine experimental embryology with genetic and molecular analyses of embryonic development at the level of individual cells in a single organism. This chapter focuses on efforts to understand the molecular and cellular events of early development in *C.elegans* with particular emphasis on events relating to the determination of embryonic cell fates. Extensive coverage of the various contributions that the study of *Caenorhabditis* has made to our knowledge of developmental biology can be found in ref. 2.

1.1 Life cycle and culture

C.elegans is a small free-living nematode with a simple anatomy (*Figure 1*). The adults are about 1 mm in length and 70 μ in diameter. Strains can be maintained in the lab on agar plates spread with a slow growing strain of *Escherichia coli* (3) or can be grown in liquid culture, and large populations of relatively synchronous worms can be obtained. Early larval stages survive freezing and so strains can be stored under

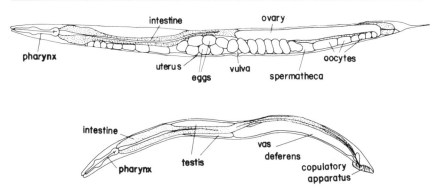

Figure 1. *Caenorhabditis elegans* hermaphrodite (above) and male (below), indicating some anatomical features. Reprinted with permission from ref. 29. Adapted from a drawing by Sulston and Horvitz (79).

liquid nitrogen for many years. *C.elegans* propagates primarily via hermaphroditic self-fertilization, although males occur spontaneously in hermaphrodite populations at a frequency of about 1/700 (3) and are able to mate with hermaphrodites. Eggs are fertilized internally and take less than a day to complete embryogenesis. Newly hatched first-stage larvae are morpholgically similar to adults with the exception of sexual characteristics. Under normal conditions, over the course of about 3 days, the larvae undergo four larval molts to become sexually mature adults (larval stages are designated L1 through L4). Under conditions of overcrowding and starvation, the L2 larvae can molt to an alternate morphological form, the dauer larva, which is long-lived and resistant to harsh environmental conditions (4,5). When conditions become more favorable, dauer larvae resume the life cycle by molting to become L4 larvae.

1.2 Overview of embryogenesis

Embryogenesis in *C.elegans* is rapid, taking about 15 h at 20°C. Because both the body wall of the mother and the eggshell are transparent, it is possible to observe the entirety of embryogenesis in live embryos under a compound microscope (6–8). This fact, combined with the resolving power of differential interference contrast microscopy, has made it possible to describe the complete embryonic cell lineage (7,8). By following individual nuclei in live animals, and by correlating these observations with serial section reconstructions of embryos and larvae, it has been possible to assign to each blastomere a descriptive name and a fate (either in terms of subsequent division pattern or terminal differentiated state).

The early cleavages follow an invariant pattern characterized by asynchrony and asymmetry. A series of asymmetric divisions during the first four cleavages result in the generation of six 'founder' cells (*Figure 2*).

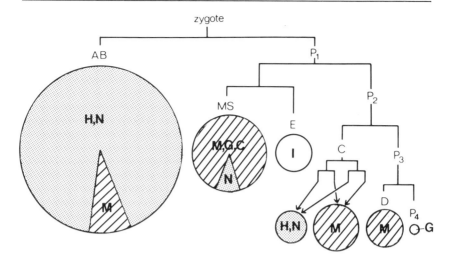

Figure 2. Generation of the founder cells and a summary of cell types derived from them. Areas of circles and sectors are proportional to the number of cells. Stippling represents typically ectodermal tissue and striping represents typically mesodermal tissue. Reprinted with permission from ref. 8. H, hypodermis; N, neurons; M, muscle; G, glands; C, coelomocytes; I, intestine; G, germ line.

Clones of cells from each of the founders behave as developmental units during early embryogenesis; each clone has a characteristic cleavage rate and develops in a characteristic way (8).

In describing the lineage, blastomeres are named for the founder cell from which they arose and are designated by the orientation of their division axis relative to the embryonic axes (anterior – posterior; left – right; dorsal – ventral): ABa is the anterior daughter of founder cell AB, ABp is the posterior daughter. ABp will give rise to the left – right pair, ABpl and ABpr.

Gastrulation begins at the 28-cell stage (100 min after the first cleavage) and is complete by the 400-cell stage (300 min). During this period the embryo remains spheroid in shape as the basic organ systems (e.g. intestine, hypodermis, pharynx) become organized and cellular differentiation begins. Cell proliferation is essentially complete by about 350 min. The remainder of embryogenesis consists of continued cell differentiation, including programmed cell death, and morphogenesis of the spheroid embryo into a worm.

The cell lineage of the embryo has provided several significant insights into embryonic development, two of which are especially relevant to a consideration of early embryogenesis. First, the embryonic lineage is essentially invariant. The pattern of cell divisions and cell fates is virtually identical in every individual. Second, as illustrated in *Figure 2*, the founder cells do not correspond exactly to specific germ layers as was believed by early nematologists (8). Three founder cells, however, do give rise to 'pure' clones: E is the sole founder of the entire intestine, D produces only body wall muscle, but not all of the body wall muscle, and P4 is the sole founder of the entire germ line. Although the complete description of the cell lineage has provided important insights into the developmental process, its greatest value is as a framework for meaningful interpretation of experimental analysis.

1.3 Genetic analysis

The *C.elegans* genome is small, with 8×10^7 bp (~1/2 that of *Drosophila*) (9) distributed over six chromosomes (10). Estimates for the total number of essential genes in *C.elegans* range from about 2000 to 3500 (3,11,12). The genetics of *C.elegans* has been recently reviewed by Herman (13). The ratio of X chromosomes to autosomes determines sex (14), with normal hermaphrodites having two X chromosomes and males having one (10). Genetic crosses are efficient because sperm produced by males are used preferentially in fertilization (15). Mutations are induced by chemical mutagenesis (3), ionizing radiation (16), or ^{32}P decay (17). Recovery of recessive mutations is aided by the hermaphrodite mode of reproduction; homozygous mutant worms appear in the F2 generation after mutagenesis with no need for sibling crosses. Consequently, large numbers of morphological and behavioral mutations have been isolated (3,18). Chromosomal rearrangements with crossover-suppressing properties have been identified (16,19–22) and used in isolation and analysis of non-conditional lethal mutations (e.g. 11,12,19,23,24). In all, nearly 800 genes have been identified and placed on the genetic map (18). Analysis of mutations is facilitated by a large collection of duplications and deficiencies (18) and by the identification of nonsense suppressors (25–27). Genetic analysis of embryogenesis has been discussed in detail (28,29) and is reviewed briefly in Section 3.

1.4 Tools for molecular analysis in *C.elegans*

Molecular analysis in *C.elegans* is aided by the worm's small genome (9), but is limited by the difficulty of obtaining usable quantities of isolated tissues and carefully staged embryos. However, three powerful tools are being developed that have greatly enhanced the potential to apply molecular analysis to mutationally defined genes in *C.elegans*.

1.4.1 A physical map of the C.elegans genome

One powerful tool enabling the application of molecular analysis in *C.elegans* is the ongoing construction of a physical map of the entire *C.elegans* genome (30). In the initial stages of construction of this map, restriction digest 'fingerprints' of randomly cloned DNA sequences were analyzed by computer to organize individual clones into many contiguous stretches of overlapping sequence (contigs). The next step is to join together these isolated contigs and correlate them with the genetic map. The ideal goal is to end with six contigs, each representing one of the linkage groups. A major effort by the entire community of *C.elegans* workers is currently in progress to correlate the isolated contigs with the genetic map. Many contigs have already been positioned and a partial physical map is taking shape (18). As the map grows, gaps between adjacent contigs are being closed by 'walking' across the gaps.

Even though the physical map may not be complete for several years, the partial map is an extremely powerful tool for isolating DNA from newly identified genes. Suppose, for example, that a newly identified embryonic lethal mutation were located within a tenth of a map unit of the myosin-encoding gene *unc-54*. It is likely that the gene identified by the embryonic lethal mutation is in the contig identified by the myosin gene sequence and is therefore already cloned. All that remains is to identify the specific DNA sequence containing the gene (see Section 1.4.3).

1.4.2 Transposon tagging and transposon mapping

Many genes identified by mutation are not closely linked to physically mapped genetic markers. In such cases, a second tool for molecular biology, transposon tagging, can be used. This requires the isolation of 'spontaneous' mutations with the phenotype of interest that arise in strains with a high frequency of transposition (31,32). The expectation is that most such 'spontaneous' mutations are the result of transposition into the gene of interest. Using DNA complementary to known transposons as a probe against restriction endonuclease-digested DNA made from the mutant strain, it is possible to identify the DNA fragment containing a novel transposon. After genetic analysis to confirm that the novel transposon co-segregates with the mutant phenotype, a DNA clone containing the novel transposon and sequences flanking it is isolated. The probability is good that the flanking DNA will be within or near the gene of interest.

A variation of transposon tagging takes advantage of differences in the number and distribution of the transposon Tc1 in different strains of *C.elegans* (33). Tc1 elements closely linked to the gene of interest can be mapped via a series of backcrosses followed by three factor crosses. Because Tc1 DNA has been cloned, it is then possible to isolate the copy of Tc1 (along with unique flanking sequence) that is most closely linked to the gene. The unique sequence can then be positioned on the physical map. Ideally, by cloning unique sequences that flank the gene on both sides, the region of the physical map containing the gene of interest can be identified.

1.4.3 Germ line transformation

The third powerful tool is genetic transformation. Nucleic acids can be introduced into nematodes by injection into the gonad or into unfertilized oocytes (34 – 36). DNA injected in this way can be transmitted to progeny as large semi-stable extrachromosomal tandem arrays or can be stably integrated into the chromosomes at random sites. Most DNAs introduced in this way exhibit appropriate spatial and temporal expression.

At present, transformation is most valuable as a tool for molecular cloning of genes identified by mutation. The most convincing argument that a DNA sequence encodes a gene of interest is the ability of the sequence to rescue a mutation in that gene. Thus, in the example given in Section 1.4.1, the DNA sequence encoding the gene identified by the embryonic lethal mutation will be in the portion of the unc-54 contig that can rescue the lethal mutation in transformation experiments.

The combination of the physical map, transposon tagging and genetic transformation should make it possible to carry out molecular analysis of virtually any gene that can be identified by mutation, and may make it possible to go in the reverse direction, targeting genes cloned by more standard means (such as cross-species DNA – DNA hybridization) for mutational analysis.

1.5 Determination of cell fates in *C.elegans*: a focus for analysis

Much interest is currently focused on specification of embryonic cell fates. What is the basis for the invariant cell lineage? Are cell fates specified cell autonomously or by positional information? What role do localized cytoplasmic determinants play in specification of cell fates? What are the determinants? How are they localized? How are embryonic asymmetries generated? What role does intercellular communication play in specification of fate? What are the mechanisms by which cytoplasmic determination and intercellular communication act? The remainder of the chapter will review early embryogenesis of *C.elegans* and report on the progress being made by studies addressing these and other related questions.

2. Gametogenesis and early events of embryogenesis

2.1 Gonadogenesis and gametogenesis

Development of the reproductive system of *C.elegans* has been described by Hirsh and co-workers (6) and Kimble and Hirsh (37). At hatching, the gonad consists of a four-cell primordium with two germ cells and two somatic cells. Both sets of cells proliferate during larval growth and ultimately form (in hermaphrodites) a two-armed bilaterally symmetric gonad (*Figure 1*) that produces sperm during late larval stages and then switches over completely to producing oocytes. The sperm that are produced migrate, or are pushed by the first oocyte, to the spermatheca, where they remain until used (15). Males have a single gonad that produces only sperm. Sperm that are introduced by mating migrate to the spermatheca (15). Unmated hermaphrodites produce about 300 embryos each; mated hermaphrodites can produce many more. For a comprehensive review of gametogenesis, see ref. 38.

The process of oogenesis begins in the distal tip of the gonad (the vulva marks the proximal gonad) where germ nuclei divide mitotically under the influence of the somatic distal tip cell (39). As the nuclei move more proximally and away from the distal tip cell, they enter meiotic prophase (6). The germ nuclei during this period exist in a syncitium with nuclei on the periphery of a central core of cytoplasm. Near the bend of the gonad, individual nuclei are incorporated into oocytes as they and a portion of the core cytoplasm are surrounded by membranes. The oocytes grow as they move towards the spermatheca. As the oocytes mature, the centrally placed oocyte nucleus, arrested in diakinesis, usually becomes positioned asymmetrically toward the distal end of the gonad.

If sperm are present in the spermatheca, the oocyte nuclear envelope breaks down and contractions of the ovarian tissue force the egg into the spermatheca where it is fertilized. However, if no sperm are present, hermaphrodites arrest oogenesis and accumulate diakinesis-stage oocytes (15). Occasionally, unfertilized oocytes pass through the spermatheca. These unfertilized oocytes do not make an eggshell, undergo at least one reductional division followed by several rounds of replication of the maternal chromosomes, but do not undergo mitosis (15).

2.2 Meiosis

Early embryogenesis has been described in some detail using light microscopy of live embryos (6,15,40,41). The timing of events occurring during the first cell cycle is very similar among individuals but the rates differ with temperature. Therefore, developmental times will be indicated by decimal fractions of the period between fertilization (0.0) and the first cleavage (1.0). The newly fertilized egg remains in the spermatheca for a brief period, during which the eggshell begins to form, saltatory motions increase in the cytoplasm, and the first meiotic division begins (15). Little is known about composition and formation of egg coverings, or about the process of embryonic activation in *C.elegans*. At about 0.3 of the first cell cycle, the first polar body buds off and becomes attached to the eggshell; it does not undergo a second division. At about 0.6, the second polar body is extruded but does not separate from the plasma membrane. In most cases, the position of the polar bodies marks the future anterior pole of the embryo. However, rare embryos with posteriorly placed polar bodies develop normally and indicate that the position of the sperm either determines or is a reliable reflection of embryonic polarity (42). The meiotic spindle is a barrel-shaped structure with no centriole or astral microtubule arrays (42). In contrast, the first mitotic spindle has a more typical spindle structure.

2.3 Pseudocleavage and pronuclear migration

The interval between 0.6 and 0.85 of the first cell cycle is an extremely critical period during which major embryonic reorganization takes place. During this period, cytoplasmic components become asymmetric in their distribution (see Section 5), cortical contractions occur, and the pronuclei migrate and join. Treatments or mutations that perturb events during this period have irreversible and dramatic effects on subsequent development.

The extrusion of the second polar body at 0.6 of the first cell cycle is accompanied by irregular contractions of the anterior cortex that continue as the egg and sperm pronuclei decondense and DNA synthesis begins (43). As the pronuclei swell, the anterior contractions subside and are gradually replaed by a single extensive cortical contraction midway between the anterior and posterior poles.

By about 0.7, this contraction becomes an almost complete furrow (called the pseudocleavage furrow) with only a narrow channel of cytoplasm connecting the anterior and posterior half of the embryo (*Figure 3a*). As the furrow forms, and while it is in place, cytoplasmic streaming of materials from the anterior to the posterior occurs (40). The maternal nucleus appears to be swept along with this flow and moves through the furrow (*Figure 3b*). As this happens the furrow relaxes and the anterior cortex contracts toward the posterior. During this period, the sperm centrosomes begin to form and the sperm pronucleus moves away from the posterior cortex. (The egg pronucleus has no centrosomes.) Once beyond the furrow, the maternal pronucleus rapidly traverses the remaining distance to meet with the sperm pronucleus in the posterior (*Figure 3c*). The pronuclear membranes remain intact and pronuclei do not fuse at this time. Pronuclear membranes break down at prometaphase, but it is possible that mixing of the parental chromosome sets does not take place until the formation of the two-cell stage nuclei.

Genetic analysis, discussed in Section 3, has identified mutations in several genes that affect early embryogenesis. Embryos from mothers homozygous for mutations in one of these genes, *zyg-11*, arrest at metaphase of meiosis II, do not form a pseudocleavage furrow, undergo exaggerated and multidirectional cytoplasmic streaming, and form multiple pronuclei at random positions in the cell after a three- to four-fold delay relative to normal; subsequent cleavages are very abnormal (44). The *zyg-11* gene product, therefore, may be necessary for initiating, or permitting the ocurrence of, cytoplasmic reorganization.

As the centrosomes continue to grow, the pronuclei move together to the center of the egg (centrate), rotate 90° and begin mitotic metaphase (*Figure 3d*). Albertson has examined centrosome behavior during the first cell cycle and has suggested that astral microtubules may be directing rotation and centration of the pronuclei (42). This hypothesis is supported by observations of the phenotype of maternal-effect lethal mutations in the gene *zyg-9* (42,44,45). Centrosomes in embryos from *zyg-9* mutant

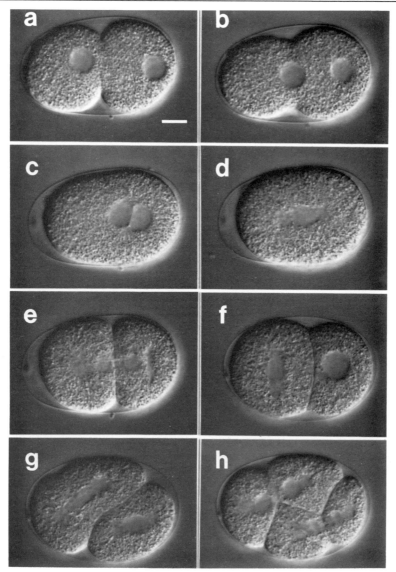

Figure 3. Nomarski micrographs of developing *C.elegans* embryos. Anterior is to the left. (**a**) Pseudocleavage stage embryo showing the maternal pronucleus early in its migration. (**b**) The maternal pronucleus is passing through the pseudocleavage furrow and the furrow is beginning to relax. (**c**) The pronuclei meet in the posterior. (**d**) The metaphase spindle is oriented along the anterior – posterior axis. (**e**) Early two-cell stage showing the differences in size of the two blastomeres, and the unusual shape of the posterior aster. (**f**) The anterior cell (AB) begins to divide. The spindle is perpendicular to the anterior – posterior axis. (**g**) The spindle of the AB is skewed in an anterior – posterior direction. The spindle in the posterior cell (P1) is aligned along the anterior – posterior axis. (**h**) The four-cell embryo. The embryo now has dorsal – ventral polarity, with the lower blastomere, EMS, the posterior daughter of P1, defining the ventral side. (Scale bar = 10 μm.)

mothers contain short microtubules and neither centrate nor rotate. Migration of the maternal pronucleus fails as well. Further support comes from experiments showing that pronuclear joining, centration and rotation are all blocked by treatment of one-cell embryos with drugs that destabilize microtubules (46).

2.4 Early cleavages

The first mitotic spindle forms initially near the center of the embryo, migrates posteriorly during prometaphase (47), then elongates more toward the posterior than toward the anterior during anaphase (42,47). Elongation is accompanied by five or six side-to-side movements of the posterior aster; the anterior aster is somewhat larger and remains fixed. The net result is that the spindle is displaced toward the posterior pole and the subsequent cleavage is, therefore, unequal, giving a large anterior cell (AB) and a small posterior cell (P1) (*Figure 3e*). By telophase, the posterior centrosome exhibits a markedly different morphology, appearing as a disc under Nomarski optics (*Figure 3e*) and as an expanding ovoid ring when stained with anti-tubulin antibodies (48).

The two blastomeres behave differently at the next cleavage. The AB cell cleaves about a minute or two before P1 and divides longitudinally (meridionally) and equally (*Figure 3f* and *h*). The P1 divides transversely (equatorially) and unequally (*Figure 3g* and *h*). The cells clonally derived from the founder cell, AB, exhibit synchronous orthogonal cleavages for the next several rounds of division (49). The unequal P1 division occurs along the same axis as the first division, to form a large anterior daughter (EMS) and a small posterior daughter (P2). These blastomeres both divide unequally giving rise to three founders (E, MS and C) which go on to divide equally and orthogonally, and P3 which divides unequally on the same axis as its mother to give rise to the founder cells D and P4. Observations of isolated blastomeres indicate that the asymmetric divisions are always oriented along the original anterior–posterior axis, but that the physical constraints of the eggshell and the presence of other blastomeres produces the more complex pattern seen in the intact embryo (49). In addition, there is an apparent reversal of polarity in P2 and P3 in that the posterior, rather than the anterior, daughter is larger and has division characteristics of a somatic founder cell (50).

The mechanisms responsible for this invariant cleavage pattern are under investigation. Because generation of the unequal first cleavage is sensitive to cytochalasin treatment, it is likely that microfilaments are required for the asymmetric positioning of the cleavage spindle (46). The critical period for this sensitivity occurs between 0.7 and 0.85 of the cell cycle when the pronuclei are migrating (51). This observation is intriguing because the spindle does not form until after the sensitive period, and is insensitive to treatment during its migration and elongation. Thus microfilaments appear to be required to establish polar spatial cues that, once established, are microfilament-independent (51). Treatment with cytochalasin during the critical period also prevents the formation of the pseudocleavage, alters the localization of germ line-specific P granules (see Section 5.1), and causes the pronuclei to meet in the center of the cell rather than in the posterior. Orientation of the spindle along the anterior – posterior axis, however, is not affected by cytochalasin treatment (46,48,51).

Subsequent cleavages appear to be of two basic types. Founder cells and their daughters divide equally and orthogonally; cells that divide to generate founder cells divide unequally along the anterior – posterior axis (48,49). By following the movement of the centrosomes in living embryos, Hyman and White have shown that following cleavage, the centrosomes divide and migrate around the nucleus to positions equidistant from the start of migration and orthogonal to the plane of the previous spindle (48). In founder cell clones no further movements of centrosomes occur and an orthogonal cleavage pattern results. However, in cells destined for unequal (or determinative) cleavages, the centrosomes, along with the nucleus to which they are now fixed, rotate 90° so that the plane of the spindle is parallel to the long axis of the egg. The motive force for the rotation seems to reside in or be dependent upon the microtubules of the astral array, since nocodazol and taxol, drugs that cause shortening of the astral microtubules, prevent the rotation (48). Also, in contrast to the rotation that occurs during the first cleavage, the later centrosome rotations are blocked by cytochalasin, implicating microfilaments in the process. The centrosome movements appear to be aligning the spindle along a pre-defined axis, indicating that there is an intrinsic polarity in cells undergoing determinative cleavages.

The relative timing of cleavages also seems to reflect an intrinsic polarity in early blastomeres. The smaller daughters from determinative cleavages tend to divide more slowly and, in general, posterior founder cell clones divide more slowly than anterior daughters (7). The results of a series of experiments using laser microsurgery has led to the suggestion that differential partitioning of cytoplasmic factors is a likely basis for the differences in cell cycle rate (52 – 54). Cytoplasts produced from early blastomeres exhibit cycles of contraction and relaxation that correspond to the cell cycles of the blastomeres from which they were obtained; cell cycle rates are not significantly affected by altering the nuclear to cytoplasmic ratio; and addition of cytoplasm from rapidly cleaving to more slowly cleaving cells tends to accelerate cleavage rates.

Some of the genes responsible for this early cleavage pattern have been identified. Strict maternal-effect lethal mutations in five genes, called *par* genes for *par*titioning defective, alter the cleavage pattern in early embryos without causing detectable defects in mitosis or cytokinesis (47; D.G.Morton and K.J.Kemphues, unpublished results). These genes are also required for other early embryonic events which are discussed

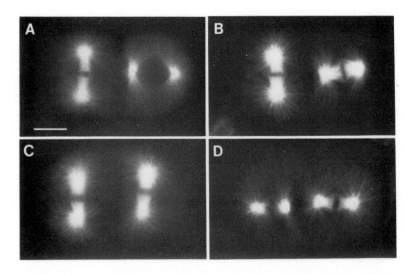

Figure 4. Indirect immunofluorescence micrographs of *C.elegans* two-cell embryos showing tubulin distribution. (**A**) Wild-type two-cell embryo, showing orthogonal spindle orientation and asynchronous mitoses. (**B**) *par-1* mutant embryo with normal spindle orientation but synchronous mitoses. (**C**) *par-2* mutant embryo showing abnormal orientation of the P1 spindle. (**D**) *par-3* mutant embryo showing abnormal orientation of the AB spindle. (Scale bar = 10 μm.) Reprinted with permission from ref. 47.

in Section 6. Embryos from mothers homozygous for mutations in *par-1*, *par-2*, *par-3* and *par-5* exhibit an equal first cleavage because the first division spindle fails to migrate. These embryos, along with embryos from *par-4* mutant mothers (which have a normal first cleavage), undergo nearly synchronous subsequent divisions.

The orientation of the subsequent divisions in embryos from *par* mutant mothers deviates from the wild-type pattern. This is most clearly seen at the second cleavage (*Figure 4*). In embryos from *par-1*, *par-2* and *par-4* mothers, orientation of the AB is always orthogonal, as it is in wild-type embryos. In *par-1* and *par-4* embryos the P1 spindle is usually parallel to the long axis but 25% of the time in *par-1*, and 20% of the time in *par-4*, the rotation of the P1 spindle fails to occur and P1 divides orthogonally. In *par-2* embryos, AB cleaves normally, but in nearly all of the embryos the P1 cleaves orthogonally. In *par-3* embryos, the spindles in both daughter cells are frequently parallel to the long axis of the embryo, like the wild-type P1. Embryos from *par-5* mutant mothers show cleavage defects intermediate between those of *par-2* and *par-3* embryos. Molecular analysis of the products of these genes should provide information about the mechanisms that control the early cleavage pattern.

3. Genetic analysis of embryogenesis

Genetic analysis of embryogenesis in *C.elegans* has two goals. The first is to learn the relative contributions of the maternal and embryonic genomes to embryonic development. The second is to understand the mechanistic basis for specific developmental processes by isolating and studying mutations that affect those processes. Of course, achieving these goals is dependent upon being able to isolate, maintain, and characterize embryonic lethal mutations.

3.1 Two classes of embryonic lethal mutations

An important aspect of embryonic development is that expression of two separate genomes is necessary to complete the process: the maternal genome, which produces the oocyte, and the embryonic genome itself. As a result, mutations in genes expressd by either or both genomes can lead to embryonic lethality, but fall into two genetically distinguishable classes: maternal-effect lethal mutations, and non-maternal lethal mutations (often referred to as 'zygotic' lethal mutations).

For non-maternal lethal mutations the genotype of the embryo determines its phenotype. Homozygous progeny (m/m) from parents heterozygous for non-maternal lethal mutations (m/+) die as embryos. In general, genes identified by non-maternal lethal mutations must be transcribed by the embryonic genome to permit successful embryogenesis.

In contrast, for maternal-effect mutations, it is the genotype of the mother that determines the phenotype of the embryo. Homozygous progeny (m/m) from parents heterozygous for a maternal-effect lethal mutation (m/+) are themselves viable but are sterile because they produce inviable offspring. Maternal-effect mutations fall into two categories. If heterozygous progeny (m/+) from a homozygous (m/m) mother (obtained by mating +/+ males with m/m mothers) are inviable, the mutation is referred to as a strict maternal-effect lethal mutation. If the m/+ progeny live, the mutation is called a partial maternal-effect lethal. Maternal-effect lethal mutations identify genes for which maternal expression is sufficient to allow completion of embryogenesis. Strict maternal-effect mutations define genes for which embryonic transcription either does not occur or is insufficient to overcome the maternal deficiency.

3.2 Isolation of embryonic lethal mutations

Three approaches have been used to identify embryonic lethal mutations in *C.elegans*:
(i)   screens for temperature-sensitive embryonic lethal mutations (45, 55 – 57);
(ii) screens for non-conditional lethal and maternal-effect lethal mutations isolated in *trans* to balancer chromosomes (e.g. 11,23,24);

(iii)   screens for maternal-effect lethal mutations identified as suppressors
of adult lethality conferred by egg-laying defective mutations
(47,49,58).

Most analysis has been carried out on genes identified by recessive
temperature-sensitive mutations, largely because their conditional nature
makes them easier to study, but also because temperature shift
experiments can provide information about the time of gene action (55).
To obtain temperature-sensitive mutations, cultures are initiated from
individual F2 progeny of mutagenized hermaphrodites at the permissive
temperature (16°C). In the next generation, a sample of their progeny
is tested at the restrictive temperature (25°C). Worms homozygous for
temperature-sensitive embryonic lethal mutations will produce dead
embryos. These strains are recovered from the 16°C culture and
maintained at the permissive temperature and can be studied at the
restrictive temperature. This screening protocol allows recovery of both
non-maternal and maternal-effect lethal mutations. Temperature-sensitive
embryonic lethal mutations are tested to learn about the mode of
expression of the genes, that is whether the genes identified by the
mutations are transcribed by the mother during oogenesis or by the
embryonic genome during embryogenesis or by both. In addition,
temperature shift experiments are carried out to investigate the time of
action of the genes or gene products. Recently, analysis has been extended
to dominant temperature-sensitive embryonic lethal mutations (59).

To identify non-conditional embryonic and larval lethal mutations,
hermaphrodites heterozygous for a balancer chromosome and its marked
homolog are treated with mutagen and F1 heterozygous progeny are
picked as individuals and allowed to self (19,23,24). [Similar types of
screens can be done using chromosomal duplications (11).] The absence
of F2 progeny homozygous for the marked chromosome indicates the
presence of a linked non-maternal lethal mutation. Embryonic lethals are
distinguished from larval lethals by the absence of arrested larvae among
self progeny from heterozygous animals (m/+). Because some of the
surviving F2 homozygous for the marker will be sterile, maternal-effect
lethal mutations can be recovered in the same screen. Maternal-effect
lethal mutations can be identified among other kinds of steriles (e.g.
spermatogenesis-defective mutations and gonadogenesis-defective
mutations) because homozygotes for maternal-effect lethal mutations lay
fertilized eggs that fail to hatch.

Because balancer chromosomes are not available for many regions
of the genome, a more general screen has been used to identify
non-conditional maternal-effect lethal mutations on any linkage group
(47,58). This screen makes use of mutations that prevent egg-laying by
hermaphrodites (21,60,61). Because eggs are fertilized internally, embryos
hatch inside hermaphrodites homozygous for egg-laying-defective
mutations and the progeny kill the mother, reducing her lifespan.

Sterile mutations, including maternal-effect lethal mutations, allow egg-laying-defective hermaphrodites to survive. For the screen, F1 progeny from mutagenized hermaphrodites of an egg-laying-defective strain are picked to individual plates and incubated until young F3 larvae are present. Only plates with worms homozygous for sterile mutations will contain surviving F2 adults, easily distinguished from the larval F3 due to their larger size. These mutations can be maintained in heterozygous stocks by selecting for worms that segregate long-lived progeny. Maternal-effect lethal mutations are readily identified among other types of sterile mutations because only they lead to accumulation of dead embryos with refractile eggshells.

3.3 Assessing the relative contributions of maternal and embryonic transcription

The first goal of genetic analysis has been to determine the extent and nature of maternal and embryonic contributions to embryogenesis. Because of the difficulty of obtaining large numbers of synchronous embryos and the impermeability of the *C.elegans* eggshell, little is known from direct observation about the nature of the maternal RNA population or the onset of embryonic transcription in early development. However, it appears that major amounts of mRNA are not produced until after the early cleavage stages. Hybridization of tritium-labeled polyuridylate to polyadenylated RNA is first detected in nuclei of squashed early embryos at around the time of gastrulation (90 – 125-cell stage) (62).

Most of our knowledge about the relative contributions of maternal and embryonic transcription to *C.elegans* embryogenesis comes from analysis of temperature-sensitive embryonic lethal mutations. A collection of 70 temperature-sensitive embryonic lethal mutations defining 55 genes has been analyzed (55 – 57,63). A series of genetic tests revealed that 54 of the 55 genes are expressed maternally (45,57). Since the screening protocol was not biased for maternally expressed genes, these results argue strongly that maternal transcription plays a dominant role in embryogenesis.

A brief survey of existing non-conditional non-maternal lethal mutations also implies a major role for maternal gene expression in embryogenesis. The relative importance of embryonic transcription should be reflected in the proportion of lethal mutations with embryonic lethal phases. Of 109 genes identified by lethal mutations for which lethal phases are reported (18), only nine are embryonic lethals. A likely basis for the abundance of larval lethal mutations is that most of the 'housekeeping' genes required for cell viability and proliferation are transcribed during oogenesis and the gene products stored in the oocyte in sufficient quantities to complete embryogenesis. Mutations in such genes would only have effects during larval growth after the wild-type gene product (supplied by the heterozygous mother) is depleted. Another possible explanation for the predominance of larval lethal mutations is that mutations in some genes critical for embryogenesis will perturb embryonic morphogenesis enough to prevent larval growth, but will not preclude hatching.

The observation of secondary phenotypes for most of the temperature-sensitive embryonic lethal mutations is consistent with the possibility that most maternal transcription satisfies 'housekeeping' needs. Temperature-sensitive mutants can be shifted to non-permissve temperature as newly hatched larvae. The expression of a mutant phenotype by such worms indicates a requirement for gene expression during the remainder of the life cycle. Based on such tests, it appears that 41 of the 54 maternally expressed genes are required outside of embryogenesis. In fact, many of them are required during larval growth, indicating that null mutations in such genes would probably behave as larval lethals. However, it is also possible that the secondary phenotypes indicate utilization of the same regulatory function in both embryonic and post-embryonic development.

In summary, studies of a set of temperature-sensitive embryonic lethal mutations and analysis of the existing pool of non-conditional lethal mutations in *C.elegans* indicate that most genes required for embryogenesis are expressed maternally in sufficient quantities to permit completion of embryogenesis.

3.4 Identifying genes that directly influence embryonic development

As discussed above, many of the genes identified by temperature-sensitive maternal-effect lethal mutations are likely to be constitutively expressed genes with general functions in cell viability or proliferation (28,45). Thus, a significant problem for a genetic approach focused on understanding the mechanistic basis of embryonic processes is distinguishing between embryonic lethal mutations that identify such 'housekeeping' genes and mutations that identify genes that exert a more direct influence on development. One might predict that strict maternal-effect lethal mutations would identify a class of genes specifically required for embryogenesis and that would, therefore, be more likely to encode embryonic control functions. Since 24 of the temperature-sensitive strict maternal-effect lethal mutations discussed in Section 2.5 exhibit secondary phenotypes when shifted to non-permissive temperature as L1 larvae, the prediction doesn't hold for temperature-sensitive mutations. However, the prediction might be valid for non-conditional strict maternal-effect mutations.

Unfortunately, it appears that even genes defined by non-conditional strict maternal-effect lethal mutations can encode essential functions outside of embryogenesis (64). In an analysis of 17 loci identified by 39 non-conditional strict maternal-effect lethal mutations on linkage group

II, it was found that only four loci mutated at frequencies typical of null mutations in *C.elegans* and are therefore likely to be specifically required for embryonic development. The low frequency of recovery of mutations at the other 13 loci makes it likely that most mutations at these loci have phenotypes other than maternal-effect lethality. Indeed, two of the 13 other loci had been previously identified as essential genes by larval lethal mutations (23). The results of this analysis led to the proposal that many non-conditional strict maternal-effect lethal mutations are rare mutations in genes that are required at multiple stages of the life cycle (64). A similar hypothesis was proposed after an analysis of female sterile loci on the X chromosome of *Drosophila* (65).

The problem of identifying mutations in genes with controlling roles in embryogenesis can be approached in two ways. The standard approach is to search for mutations that disrupt particular embryonic processes based on the phenotypes of embryonic lethal mutants. This approach has been fruitful in genetic studies of *Drosophila* pattern formation (see Chapters 1 and 2), largely because pattern defects are reflected in cuticular bristle configurations that are easily scored in terminal-stage embryos. It is not yet clear how beneficial analysis of terminal phenotypes will be in *C.elegans*; few genes (e.g. *glp-1*, see Section 6.2) have been singled out for detailed analysis based on their terminal embryonic phenotypes. However, a set of mutations defining genes important for early cleavages, the *par* genes, have been identified by screening for maternal-effect lethal mutations causing altered early cleavage patterns.

A second approach focuses on identifying genes that are likely to be specifically required for embryogenesis. These include genes that mutate at high frequency only to strict maternal-effect lethality ('pure' maternal genes) (64) and genes that mutate at high frequency to non-conditional embryonic (as opposed to larval) lethality. This approach has the advantages that it can be done systematically and that saturation for such genes is conceivable, but has the disadvantage that important genes will be missed since some embryonic control genes will have essential functions outside of embryogenesis as well.

4. Mechanisms of determination in *C.elegans* embryos: laser ablation experiments

One of the first uses of the embryonic lineage was to assess the extent to which the invariant pattern was determined by cell-autonomous mechanisms versus intercellular or inductive interactions. Previous studies had shown that the properties of the founder cells (cell division rate and cleavage pattern) were determined cell-autonomously (49). A more general answer to this question was obtained from an extensive series of blastomere ablation experiments using a laser microbeam to kill specific embryonic cells after the 50-cell stage (8). In all but two cases cells behaved in a completely cell-autonomous fashion: the cells that would normally arise from the ablated cells were missing, and adjacent cells did not regulate to replace the missing cells. In the two cases where regulation did occur, no additional cell divisions took place. Rather, the fate of one of the daughters of the ablated cell was assumed by an adjacent cell from a different lineage. The normal fate of the replacing cell was not expressed. Even with the two exceptions, these experiments indicate that developmental fate for most embryonic cells beyond the 50-cell stage is determined cell-autonomously.

Unfortunately, interpretation of laser ablation experiments on embryos prior to the 50-cell stage is difficult, so the question of cell autonomy has not been resolved for many of the early blastomeres (8). One pair of early blastomeres that has been examined, however, has proven to be an important exception to the general rule of cell-autonomous determination in *C.elegans* embryos and is discussed in detail in Section 6.

5. Determination by cytoplasmic localization

5.1 Asymmetry and localization in early embryos

The cell-autonomous generation of non-identical daughters by cleavage implies some form of asymmetric partitioning of cellular components, either nuclear or cytoplasmic. One possible nuclear mechanism for generating early embryonic asymmetries, 'chromosome imprinting' with differential segregation of maternal and paternal chromosomes, has been ruled out (66) but other forms of asymmetric chromosome marking and distribution cannot be excluded. Most evidence, however, indicates that cytoplasmic, rather than nuclear, asymmetries play the major role in determining early embryonic cell fates.

5.1.1 Polarity

The early embryo has a clear anterior – posterior polarity with the posterior pole marked by the position of the sperm nucleus. No evidence of dorsal – ventral polarity is present in oocytes or one-cell embryos. The egg nucleus and polar bodies usually mark the anterior end, but are occasionally observed in the posterior with no detrimental effect on development (42). It has not been possible to ascertain the relationship between polarity exhibited by the oocyte and embryonic polarity.

In early zygotes, while the meiotic divisions are taking place and the eggshell is forming (from 0.0 to 0.6 of the first cell cycle), visible cytoplasmic components are distributed uniformly and the only indication of polarity is given by the positions of the nuclei. However, during the period from 0.6 to 0.85 there is a major reorganization of the zygote cytoplasm that leads to striking cytoplasmic asymmetries.

5.1.2 *Asymmetric distribution of microfilaments in one-cell embryos*

Microfilaments can be detected in fixed *C.elegans* embryos using rhodamine-labeled phalloidin (67). At early cleavage stages, a cortical network of microfilaments is readily detected. Interspersed among and slightly anterior to the filaments of the cortical network are a large number of brightly staining foci. In early embryos (0.0 – 0.6) the intensity of staining is uniform over the entire cortex. During the period from 0.6 to 0.7, embryos show the same overall pattern but the anterior 40% of some embryos is more intensely stained, due apparently to a higher density of foci. By 0.85, the time of pronuclear meeting, all embryos exhibit this asymmetric pattern of actin staining. The asymmetric pattern persists through metaphase of the first cleavage in most embryos, but is only detected rarely in two-cell embryos. The microfilament pattern is disrupted in the presence of cytochalasin, but recovers rapidly after the drug is washed out (51).

5.1.3 *Asymmetric distribution of germ line-specific granules*

During the same period that microfilaments become concentrated in the anterior cortex, germ line-specific granules (P granules) become localized to the posterior cortex. P granules are cytoplasmic particles that segregate with germ line (P) cells and are detected by indirect immunofluorescence microscopy (68). The P granules are probably identical to a set of similarly distributed particles that have been detected by electron microscopy and appear analogous to germ line-specific granules in other organisms (69). Although the behavior of P granules is consistent with their playing a role in specification of the germ line, neither the composition nor function of the granules is yet known. However, their discovery has made it possible to study the process of asymmetric cytoplasmic partitioning in detail.

In mitotic germ cells and early oocytes, P granules are associated with the nuclear envelope. From late in oogenesis to about 0.7 of the first cell cycle, however, they are uniformly distributed throughout the cytoplasm (*Figure 5a*). During the cytoplasmic reorganization, when the pronuclei are migrating, the P granules become localized to the posterior cortex (*Figure 5c*). The basis of this localization seems to be redistribution of the existing granules, but localized degradation and *de novo* synthesis of the granules may also be involved (46). At the first cleavage, the granules are segregated specifically to the P1 cell. They subsequently localize to the posterior cortex (*Figure 5e*) and are segregated specifically into the P2 (*Figure 5g*). Asymmetric partitioning of P granules occurs again at the division of P2 and P3. The granules are distributed symmetrically at the P4 and subsequent germ line divisions.

The localization of the granules in the first cell cycle is dependent upon microfilaments but not upon microtubules. P granules localize normally

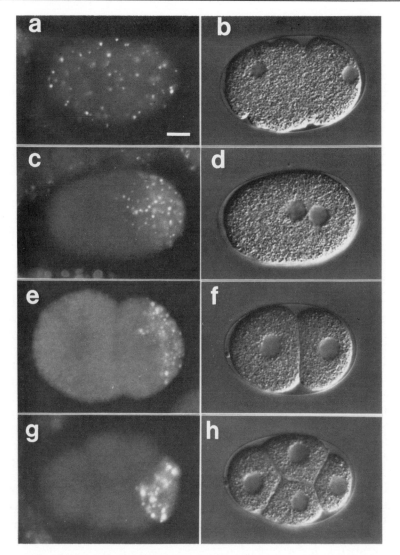

Figure 5. P granule localization in *C.elegans*. Panels (**a**), (**c**), (**e**) and (**g**): indirect immunofluorescence micrographs showing P granule distribution. Panels (**b**), (**c**), (**f**) and (**h**): Nomarski micrographs. Paired panels are different embryos of comparable developmental stages. (Scale bar = 10 μm.)

in permeabilized one-cell embryos exposed to a variety of microtubule inhibitors, but do not in embryos of similar age treated with cytochalasin B or D (46). In embryos allowed to develop in the presence of cytochalasin, the P granules become localized around the pronuclei in the center of the cell. Short pulses of cytochalasin during the first cell cycle indicate that the sensitivity of P granule localization is limited to the brief period between 0.7 and 0.85 of the first cell cycle (51). Eight-minute pulses during this period not only lead to disappearance of the pseudocleavage, meeting of pronuclei in the center of the egg, and blockage of the asymmetric placement of the first cleavage spindle as described in Section 4.3, but also prevent the posterior localization of the P granules. Pulses before this period do not prevent the localization and pulses after this period do not alter the position of the already localized granules.

The relationship between the localization of microfilaments to the anterior and localization of P granules to the posterior is not clear. Strome (67) has proposed that the actin foci represent part of a contractile network which, when concentrated in the anterior, leads to contractions that push cytoplasmic components, including P granules, to the posterior where the granules become attached to a cytochalasin-resistant structure. However, in more recent experiments, in some embryos that had recovered from a cytochalasin pulse before the critical period, both microfilaments and P granules were localized in the posterior (51). Thus, anterior localization of microfilaments does not seem to be required for posterior P granule localization.

5.2 Cytoplasmic localization in the determination of the gut lineage

Although it has not been possible to investigate directly the mechanism of determination in the germ line, it has been possible to carry out informative experimental analysis of the intestinal lineage. Two features of intestinal development make it especially favorable for experimentation. First, the intestine arises exclusively and entirely from the E founder cell which is generated at the third cleavage (*Figure 2*). Second, differentiated intestinal cells produce intestine-specific 'gut granules' that can be visualized with polarization (70) or fluorescence optics (71) in live cells and produce an intestine-specific esterase that can be stained in fixed animals (72). Because the appearance of gut granules and esterase requires embryonic transcription (43,72), production of these markers by an embryo is taken as an indication that intestine determination and differentiation has taken place.

Evidence for cell-autonomous determination of intestinal fate comes from blastomere isolation and cleavage block experiments. Isolated P1, EMS and E blastomeres produce partial embryos with differentiated intestinal cells, while other isolated blastomeres do not (49,72). In cleavage block experiments, embryos that are permeabilized in the presence of microfilament inhibitors, microtubule inhibitors, or both, fail to cleave, but continue DNA synthesis. After incubation for a period of time equal to normal embryogenesis, treated embryos can be assayed for the production of a particular marker of differentiation (e.g. gut granules). Using different permeabilization techniques and culture media, early embryos have been cleavage-blocked at various developmental stages and assayed for the production of gut granules (49,72,73). Only cells in the lineage leading to the production of the E cell produce gut granules (P1, EMS, E and E progeny cells, see *Figure 2*). Cleavage-blocked one-cell embryos do not produce gut granules. Results consistent with these have been obtained using the expression of an intestine-specific esterase as a marker for differentiation (72).

These results imply first, that determination of the gut lineage is cell-autonomous, and second, that the potential to differentiate gut is present in the P1 cell and is sequentially segregated to the E cell. To determine if the potential to differentiate intestine resides in the nucleus or cytoplasm, P1 cytoplasm was tested for the ability to transform the fate of an AB nucleus (52,74). The P1 nucleus and attached centrosomes were extruded through a hole in the eggshell, leaving a partial embryo composed of a large P1-derived cytoplast and an intact AB. When the P1 cytoplast was fused with one of the AB granddaughters, the resulting hybrid cell cleaved to produce many daughters. At least some of these daughters produced intestine-specific granules. This positive result indicates that P1 cytoplasm carries factors specifying differentiation of intestine.

Taken together, cleavage block, blastomere isolation and indirect cytoplasmic transfer experiments provide a strong argument that determination of the intestine is dependent upon the sequential localization of cytoplasmic factors that are present and potentially active in the cytoplasm of two-cell embryos. However, the cell-autonomous nature of determination of the E lineage has recently been questioned (50). In one series of ablation experiments, removal of the P2 blastomere affected the cell cycle rate and number of divisions of the E cell clone. None of six partial embryos that were scored produced gut granules, leading to the suggestion that E fate could be dependent upon an induction by P2. However, in another set of P2 ablations performed under different conditions, removal of P2 did *not* prevent the production of gut granules in 26 of 41 partial embryos (75). In 13 of the 15 negative embryos, P2 was removed very soon after the division of P1 when P2 and EMS are still connected by a cytoplasmic bridge. Thus the negative result could be interpreted as sensitivity to damage at this time or as evidence of an induction early in the P2 cell cycle. If the latter is true, then an event that mimics this induction can occur intracellularly in cleavage-blocked two-cell embryos, since the P1 cell in such embryos can differentiate as intestine.

It is not clear why cleavage-blocked one-cell embryos fail to produce intestinal markers. It does not appear to be due to a greater sensitivity of one-cell embryos to damage since cleavage-blocked one-cell embryos are capable of expressing a marker of hypodermal differentiation, a differentiation that occurs later in development than intestine (73). It could be indicative of a need for embryonic transcription in the two-cell embryo, but more likely reflects the need for a compartmentalization of cytoplasmic components before the intestine marker can be expressed (72).

The relationship between expression of intestinal markers and DNA synthesis has been examined using aphidicholin to block DNA synthesis at various times in early development (43). These experiments demonstrated that expression of intestinal markers is not controlled by counting the normal rounds of DNA synthesis, by reaching a critical DNA: cytoplasm ratio, or by lengthening of the cell cycle. Rather, expression of intestinal markers is dependent upon a critical round of DNA synthesis that occurs during the initial cell cycle of the E founder cell. This is consistent with a model for cytoplasmic determinants; it could be at this cell cycle that determinants migrate to the nucleus and interact with newly synthesized DNA to activate the program of gene expression leading to differentiation of intestinal cells.

The pattern of expression of markers of muscle and hypodermal differentiation is consistent with cell-autonomous determination of at least some of the cells in the muscle and hypodermal lineage (73), but direct evidence for cytoplasmic determinants is lacking. Interestingly, when cleavage-blocked two-cell embryos were scored for both intestine and hypodermal markers or for intestine and muscle markers, it was found that the two markers were never expressed simultaneously in a single cleavage-arrested cell (73).

5.3 Genetics of asymmetry and localization

Analysis of the *par* genes is providing insights into the mechanisms of cytoplasmic localization and generation of asymmetry. As previously discussed in Section 4.3, *par* mutations alter cleavage patterns in the early embryo due to defects in the asymmetric positioning and orientation of mitotic spindles and timing of divisions in the early embryo. Interestingly, the same mutations lead to defects in P granule localization and other aspects of early embryonic asymmetry.

Mutations in *par-1* and *par-4* block localization of P granules so that granules are present in all cells of the early embryo (*Figure 6B* and *F*). The granules are apparently degraded during subsequent embryogenesis since they are not detected at late stages. Mutations in *par-2, par-3* and *par-5* do not completely block P granule localization and lead to variable distributions of the granules. Many *par-2* embryos have P granules in all cells, but embryos like the one shown in *Figure 6C* are frequently seen. In this embryo, the granules apparently localized properly at the first division but failed to localize at the second division. In *par-3* mutant embryos, P granule distributions like those shown in *Figure 6D* and *E* are common. These distributions appear to arise from incomplete localization at the first cleavage followed by successful localization at the second cleavage.

The correlation between cleavage pattern defects and abnormalities in

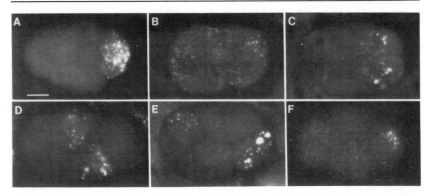

Figure 6.   P granule distribution in four-cell *C.elegans* embryos. **(A)** Wild-type embryo. **(B)** *par-1* embryo. **(C)** *par-2* embryo. **(D)** and **(E)** *par-3* embryos. **(F)** *par-4* embryo. (Scale bar = 10 μm.) Reprinted with permission from ref. 47.

P granule localization in the *par* mutants was not anticipated. None of the *par* mutations was identified on the basis of defects in P granule localization, yet all *par* mutations exhibit such defects. Because this correlation holds for mutations in five separate genes, the implication is that both cytoplasmic localization and spindle orientation are dependent upon a common system. In addition to defects in P granule localization, mutations at *par-2, par-3* and *par-5* block the transient asymmetric distribution of cortical microfilaments in one-cell embryos (M.Kusch, C.Kirby and K.J.Kemphues, unpublished results). Furthermore, pseudo-cleavage and the position of pronuclear meeting are affected in some *par* mutants (C.Kirby, B.Suh and K.J.Kemphues, unpublished results).

The *par* mutations provide circumstantial evidence for a functional link between cytoplasmic localization and the determination of at least two differentiated cell fates: intestine and germ line. Strong mutations in *par-1* and *par-4* produce embryos which arrest as amorphous masses of apparently differentiated cells. The cell lineages leading to this terminal state are clearly abnormal; *par-1* arrests with about 800 cells on average instead of the wild-type 558, and the number of differentiated cells of any given type varies in both *par-1* and *par-4* mutants. In spite of the abnormal development of these embryos, most contain some hypodermal cells, muscle cells, neurons, and undergo programmed cell deaths. However, less than 1% of *par-1* embryos and no *par-4* embryos produce intestinal cells. The particular sensitivity of intestine to these mutations indicates that *par-1* and *par-4* are required for differentiation of intestinal cells and may mean that the same system that localizes P granules also localizes determinants of intestine.

Because cytoplasmic partitioning of P granules to the germ line is defective, it seems possible that germ cell determination might also be affected by *par* mutations. It is impossible to assay for the presence of germ cells in terminal-stage *par* mutants because no markers of germ cell differentiation are known. However, evidence from leaky mutations in *par-2* and *par-3* indicates that *par* mutations may also affect the determination of germ cells.

A large proportion of the surviving progeny from mothers homozygous for leaky *par-2* and *par-3* alleles are sterile. Because the sterility is a strict maternal effect, the phenotype is called 'grandchildless' after similar mutations in *Drosophila* (76, see Chapter 1). Most sterile progeny of *par-2* and *par-3* mothers are lacking germ cells, although in a few survivors from a temperature-sensitive allele of *par-2*, germ cells are present but produce abnormal gametes. Absence of germ cells in adults is apparently due to loss or misbehavior of the germ line progenitor cells during embryogenesis, since the two precursors to the germ line cannot be detected in those surviving newly hatched L1 larvae from *par-2* and *par-3* mutant mothers that grow to be sterile adults (48, N.Cheng and K.Kemphues, unpublished results). Some weak *par-4* alleles and the single *par-5* allele also express grandchildless phenotypes.

In summary, the *par* mutations affect many aspects of asymmetry in early embryos including blastomere size, spindle orientation, division rate, microfilament distribution and P granule localization. All of these phenotypes could result from defects in a maternally encoded system for cytoplasmic localization in early embryos. The sensitivity of intestine and germ line differentiation to mutations in *par* genes is consistent with the view that this maternal system localizes cytoplasmic factors necessary for the determination of these cell types. Because the mutations make it possible to isolate the DNA corresponding to the *par* genes, it should soon be possible to determine the molecular mechanisms responsible for establishing or maintaining asymmetry.

6. Determination by intercellular interactions in early embryos

6.1 Induction in early embryos

Although the results of laser ablation experiments (Section 4) indicate that, in general, cell-autonomous mechanisms operate to establish embryonic cell fates, at least one intercellular interaction occurs and is critical in early embryonic development. In a series of ablation and micromanipulation experiments on live embryos, Priess and Thomson have shown that the first two daughters of the AB founder cell, ABa and ABp, are equivalent at birth, but express different fates depending upon their position in the four-cell embryo (75). The AB cleavage is equal and transverse, but in late anaphase the spindle skews along the anterior – posterior axis (*Figure 2*). The lineages of the AB daughters are not identical. Among other differences, progeny of ABa contribute to the pharynx while progeny of ABp do not. The 37 pharyngeal muscles can be visualized using tissue-specific antibodies and were therefore used as markers to study the basis for the differences in the lineages. ABa gives rise to 19 pharyngeal muscles in the anterior pharynx. The remaining 18 posterior pharyngeal muscles are produced by the MS founder cell. Blastomere isolation experiments indicate that the production of muscles by the ABa lineage is dependent upon an intercellular interaction with cells in the P1 lineage. After removal of P1 at the two-cell stage, an isolated AB undergoes many divisions producing differentiated cells but does not produce pharyngeal muscles. In contrast, an isolated P1 does produce cells that are similar to pharyngeal muscles.

It seems likely that cells in the EMS lineage are the source of the inductive signal. Embryos from which EMS was removed do not produce pharyngeal muscles while partial embryos missing the P2 lineage produce pharyngeal muscle in apparent excess of that expected from MS alone. However, the presence of the MS-derived pharyngeal muscles in the partial embryos complicates the interpretation of this latter experiment.

The induction of pharyngeal muscle may occur as early as the 28-cell stage. At this stage all five precursor cells are clustered on the ventral side. When all five are ablated, no pharyngeal muscles are produced. When the three AB-derived precursors are ablated only the posterior pharyngeal muscles are produced. When the two P1-derived precursors are ablated, the embryos produce only the anterior pharyngeal muscles. This result is consistent with the notion that the AB-derived precursors are already determined by the 28-cell stage. The proximity of the two P1-derived pharyngeal precursors to the AB precursors makes them the most likely source of the inducing signal, but experiments to test this possibility have not been reported.

Clearly, the ABa cells specifying pharyngeal muscles are determined by intercellular interactions. The question of whether the many other differences between the ABa and ABp lineages are also determined by induction was answered by the simple experiment of reversing the positions of ABa and ABp. By exerting external pressure against the mitotic pole of AB that is skewing in an anterior direction at anaphase, the direction of movement can be reversed, causing the daughter that would have become ABa to come to rest in the ABp position. In spite of the reversal, the embryos develop into complete larvae with normal morphology, behavior and left – right asymmetry. Since there is no major reorganization of cells in the reversed embryos, this indicates that ABa and ABp are functionally identical near the end of the cleavage that generates them, and strongly implies that all the differences in their lineages depend upon positional information. Furthermore, this result means dorsal – ventral polarity is not fixed prior to the cleavage of AB, and may mean that the dorsal – ventral axis arises randomly, depending on the direction of movement of the AB spindle.

6.2 Genetic analysis of embryonic induction

Genetic analysis of at least one component of induction in the ABa lineage is under way. Embryos from mothers homozygous for any of four maternal-effect lethal alleles of the gene *glp-1* are missing the anterior portion of the pharynx (58). Since the anterior portion of the pharynx is made up almost exclusively of cells from the ABa lineage, experiments were done to test the possibility that the phenotype could result from inductive failure. The two P1-derived pharyngeal precursor cells were ablated at the 28-cell stage in the mutant embryos. In the wild-type this ablation does not prevent the differentiation of the anterior pharynx. In the treated embryos the early cell lineage of the three ABa-derived precursors was normal and the progeny cells gastrulated properly, but the embryos produced no pharyngeal muscles. Therefore, all the pharyngeal muscles in the mutant animals must be coming from the P1-derived precursors, implying that the defect is specific to the induced cells in the AB lineage.

The phenotype of non-maternal alleles of *glp-1* indicates that the gene is required for intercellular communication between the somatic gonad and the germ line. Loss of function of the *glp-1* gene leads to sterility resulting from a failure in germ line proliferation because all the germ cells enter meiosis (77). In wild-type worms, continued proliferation of the germ line is dependent upon an interaction between the distal tip cell, part of the somatic gonad, and the germ line (39). Laser ablation of the distal tip cell has the same effect as loss of function of *glp-1*; all germ cells enter meiosis. Analysis of genetic mosaics demonstrated that the *glp-1* product is produced in the germ line, and thus *glp-1* seems to be essential for receiving, or responding to, a signal from the distal tip cell (77).

The four strict maternal-effect lethal alleles indicate that, in addition to its function in the germ line, the *glp-1* gene product is stored in the oocyte and is required for embryogenesis. Here it seems to be required for at least two functions. The first is specification of anterior pharyngeal cells as described above. In addition, it is required for proper specification or behavior of hypodermal cells. Many embryos produced by *glp-1* mutant mothers fail to undergo normal morphogenesis (58,77). A major component of morphogenesis, elongation, is dependent upon the integrity of the hypodermis (8,78). At least part of the basis for abnormal morphogenesis in *glp-1* embryos is the presence of supernumerary or misplaced hypodermal cells. The temperature-sensitive period for both pharyngeal and morphogenetic defects occurs between the four- and 28-cell cleavage stages, during the same period that the induction of pharyngeal cells is believed to occur. However, it is not clear whether the hypodermal defect is the result of an inductive failure.

By analogy with its function in the germ line, it seems likely that *glp-1* is acting as part of the receiving mechanism for an inductive signal necessary for specification of anterior pharynx. However, *glp-1* mutations do not affect all of the lineages that distinguish ABa from ABp. In normal development only ABp produces γ-aminobutyric acid-containing motor neurons in the ventral nerve cord. Since these neurons are present in *glp-1* mutant embryos, other cell – cell interactions must not be dependent upon *glp-1*.

7. Concluding remarks and future prospects

C.elegans embryogenesis has now been described in intricate detail and provides the only model system with a complete fate map at the level of individual cells. The major modes of determination have been elucidated and appear to be similar to those generally found in other systems. Cell-autonomous determination appears to be the rule, but at least one early inductive interaction plays a critical role. Embryogenesis appears to be controlled predominantly by maternal gene expression, although important contributions from embryonic expression have not been ruled out.

Significant progress has been made toward understanding mechanisms for determination in early development. Anterior – posterior polarity is reflected in or specified by the sperm entry point but is not manifested in any obvious way in the cytoplasm until 0.6 of the first cell cycle. At this time there is a major reorganization of cytoplasmic components leading to several striking cytoplasmic asymmetries. During this period, microfilaments become concentrated in the anterior cortex, P granules become localized to the posterior cortex, and spatial cues for asymmetric positioning of the first mitotic spindle are established. Presumably as a result of this reorganization, the posterior daughter cell, P1, undergoes an unequal division while the AB cell does not. The ability to undergo unequal divisions is passed on to the germ cell progenitors and perhaps to the EMS cell, but not to the other somatic founder cells.

Localized cytoplasmic factors appear to be important for determination of the intestinal lineage and may also be necessary for determination of the germ line. The localization of these factors is coordinated with a complex cleavage pattern that is controlled at the level of centrosome movements and spindle placement. The phenotypes of the *par* mutations indicate that this coordination of localization and spindle positioning is accomplished by using the same system to orient both processes.

In addition to cell-autonomous determination, determination of some cell types in the AB lineage depends on intercellular interactions. A gene, *glp-1*, has been identified that plays an important role in one of these interactions, the induction of anterior pharynx.

The prospects are good for the future of *C.elegans* embryology. In spite of the limitations of small embryos and the difficulty of obtaining large numbers of synchronous embryos, classical experimental approaches can still provide additional information about embryonic patterning and specification of cell fates at the cellular level. However, rapid progress towards a molecular understanding of these and other embryonic processes will depend on the identification and investigation of genes with controlling roles in those processes. Several important genes have already been identified. Large scale hunts for embryonic lethal mutations over the next few years should lead to the identification of many more.

8. Acknowledgements

I am grateful to Mariana Wolfner, Diane Morton, Colleen Kirby and Phil Carter for suggestions on the manuscript and to members of my laboratory for allowing me to cite their unpublished results.

9. References

1. Wilson,E.B. (1925) *The Cell in Development and Heredity*. Macmillan, New York, 3rd edition.
2. Wood,W.B. (1988) *The Nematode Caenorhabditis elegans*. Cold Spring Harbor Laboratory Press, New York.
3. Brenner,S. (1974) The genetics of *Caenorhabditis elegans. Genetics,* **77**, 71–94.
4. Cassada,R.C. and Russell,R.L. (1975) The dauerlarva, a post-embryonic developmental variant of the nematode *Caenorhabditis elegans. Dev. Biol.,* **46**, 326–342.
5. Riddle,D.L. (1988) The dauer larva. In *The Nematode Caenorhabditis elegans*. Wood,W.B. (ed.), Cold Spring Harbor Laboratory Press, New York.
6. Hirsh,D., Oppenheim,D. and Klass,M. (1976) Development of the reproductive system of *Caenorhabditis elegans. Dev. Biol.,* **49**, 200–219.
7. Deppe,U., Schierenberg,E., Cole,T., Kreig,C., Schmitt,D., Yoder,B. and von Ehrenstein,G. (1978) Cell lineages of the embryo of the nematode *Caenorhabditis elegans. Proc. Natl. Acad. Sci. USA,* **75**, 376–380.
8. Sulston,J., Schierenberg,E., White,J. and Thomson,N. (1983) The embryonic cell lineage of the nematode *Caenorhabditis elegans. Dev. Biol.,* **100**, 67–119.
9. Sulston,J. and Brenner,S. (1974) The DNA of *Caenorhabditis elegans. Genetics,* **77**, 95–104.
10. Nigon,V. (1949) Les modalites de la reproduction et le determinisme de sexe chez quelques nematodes libres. *Ann. Sci. Nat. Zool. Biol. Anim.* (ser. 11), **2**, 1–132.
11. Meneely,P. and Herman,R.K. (1979) Lethals, steriles and deficiencies in a region of the X chromosome of *Caenorhabditis elegans. Genetics,* **92**, 99–115.
12. Clark,D.V., Rogalski,T.M., Donati,L.M. and Baillie,D.L. (1988) The *unc-22 (IV)* region of *Caenorhabditis elegans:* genetic analysis of lethal mutations. *Genetics,* **119**, 345–353.
13. Herman,R.K. (1988) Genetics. In *The Nematode Caenorhabditis elegans*. Wood,W.B. (ed.), Cold Spring Harbor Laboratory Press, New York.
14. Madl,J.E. and Herman,R.K. (1979) Polyploids and sex determination in *Caenorhabditis elegans. Genetics,* **93**, 393–402.
15. Ward,S. and Carrel,J.S. (1979) Fertilization and sperm competition in the nematode *Caenorhabditis elegans. Dev. Biol.,* **73**, 304–321.
16. Herman,R.K., Albertson,D. and Brenner,S. (1976) Chromosome rearrangements in *Caenorhabditis elegans. Genetics,* **83**, 91–105.
17. Babu,P. and Brenner,S. (1981) Spectrum of ^{32}P-induced mutants of *Caenorhabditis elegans. Mutat. Res.,* **82**, 269–273.
18. Hodgkin,J., Edgley,M., Riddle,D.L., Albertson,D.G. *et al.* (1988) In *The Nematode*

Caenorhabditis elegans. Wood,W.B. (ed.), Cold Spring Harbor Laboratory Press, New York, Appendix 4.

19. Herman,R.K. (1978) Crossover suppressors and balanced recessive lethals in *Caenorhabditis elegans*. *Genetics*, **88**, 29–65.

20. Rosenbluth,R.E. and Baillie,D.L. (1981) The genetic analysis of a reciprocal translocation, *eT1 (III:V)*, in *Caenorhabditis elegans*. *Genetics*, **99**, 415–428.

21. Ferguson,E.L. and Horvitz,H.R. (1985) Identification and characterization of 22 genes that affect the vulval cell lineages of the nematode *Caenorhabditis elegans*. *Genetics*, **110**, 17–72.

22. Fodor,A. and Deak,P. (1985) The isolation and genetic analysis of a *Caenorhabditis elegans* translocation (*szT1*) strain bearing an X-chromosome balancer. *J. Genet.*, **64**, 143–167.

23. Sigurdson,D.C., Spanier,G.J. and Herman,R.K. (1984) *Caenorhabditis elegans* deficiency mapping. *Genetics*, **108**, 331–345.

24. Rosenbluth,R.E., Cuddeford,C. and Baillie,D.L. (1983) Mutagenesis in *Caenorhabditis elegans*. I. A rapid eukaryotic mutagen test system using the reciprocal translocation *eT1 (III.V)*. *Mutat. Res.*, **110**, 39–48.

25. Waterston,R.H. and Brenner,S. (1978) A suppressor mutation in the nematode acting on specific alleles of many genes. *Nature*, **275**, 715–719.

26. Waterston,R.H. (1981) A second informational suppressor, *sup-7 X* in *Caenorhabditis elegans*. *Genetics*, **97**, 307–325.

27. Hodgkin,J. (1985) Novel nematode amber suppressors. Genetics, **11**, 287–310.

28. Wilkins,A.S. (1986) *Genetic Analysis of Animal Development*. John Wiley & Sons, New York.

29. Kempues,K.J. (1988) Genetic analysis of embryogenesis in *Caenorhabditis elegans*. In *Developmental Genetics of Higher Organisms*. Malacinski,G.M. (ed.), Macmillan, New York.

30. Coulson,A., Sulston,J., Brenner S. and Karn,J. (1986) Towards a physical map of the genome of the nematode *Caenorhabditis elegans*. *Proc. Natl. Acad. Sci. USA*, **83**, 7821–7825.

31. Moerman,D.G., Benian,G.M. and Waterston,R.H. (1986) Molecular cloning of the muscle gene *unc-22* in *Caenorhabditis elegans* by Tc1 transposon tagging. *Proc. Natl. Acad. Sci. USA*, **83**, 2549–2583.

32. Collins,J., Saari,B. and Anderson,P. (1984) Activation of a transposable element in the germ line but not the soma of *Caenorhabditis elegans*. *Nature*, **328**, 726–728.

33. Emmons,S.W. (1988) The Genome. In *The Nematode Caenorhabditis elegans*. Wood,W.B. (ed.), Cold Spring Harbor Laboratory Press, New York.

34. Kimble,J., Hodgkin,J., Smith,T. and Smith,J. (1982) Suppression of an amber mutation by microinjection of suppressor tRNA in *Caenorhabditis elegans*. *Nature*, **299**, 456–458.

35. Stinchcomb,D.T., Shaw,J.E., Carr,S.H. and Hirsh,D. (1985) Extrachromosomal DNA transformation of *Caenorhabditis elegans*. *Mol. Cell. Biol.*, **5**, 3434–3496.

36. Fire,A. (1986) Integrative transformation of *Caenorhabditis elegans*. *EMBO J.*, **5**, 2673–2680.

37. Kimble,J. and Hirsh,D. (1979) The postembryonic cell lineages of the hermaphrodite and male gonads in *Caenorhabditis elegans*. *Dev. Biol.*, **70**, 396–417.

38. Kimble,J. and Ward,S. (1988) Germ-line development and fertilization. In *The Nematode Caenorhabditis elegans*. Wood,W.B. (ed.), Cold Spring Harbor Laboratory Press, New York.

39. Kimble,J.E. and White,J.G. (1981) On the control of germ cell development in *Caenorhabditis elegans*. *Dev. Biol.*, **81**, 208–219.

40. Nigon,V., Guerrier,P. and Monin,H. (1960) L'Architecture polaire de l'oeuf et movements des constituants cellulaires au cour des premieres etapes du developpement chez quelque nematodes. *Bull. Biol. Fr. Belg.*, **94**, 132–201.

41. Nigon,V. (1965) Developpement et reproduction des nematodes. In *Traite de Zoologie*. Grasse,P.P. (ed.), Masson et Cie, Paris, Volume 4.

42. Albertson,D. (1984) Formation of the first cleavage spindle in nematode embryos. *Dev. Biol.*, **101**, 61–72.

43. Edgar,L.G. and McGhee,J.D. (1988) DNA synthesis and the control of embryonic gene expression in *C.elegans*. *Cell*, **53**, 589–599.

44. Kemphues,K.J., Wolf,N., Wood,W.B. and Hirsh,D. (1986) Two loci required for cytoplasmic organization in early embryos of *Caenorhabditis elegans. Dev. Biol.,* 113, 449–460.

45. Wood,W.B., Hecht,R., Carr,S., Vanderslice,R., Wolf,N. and Hirsh,D. (1980) Parental effects and phenotypic characterization of mutations that affect early development in *Caenorhabditis elegans. Dev. Biol.,* 74, 446–469.

46. Strome,S. and Wood,W.B. (1983) Generation of asymmetry and segregation of germ-line granules in early *C.elegans* embryos. *Cell,* 35, 15–25.

47. Kemphues,K.J., Priess,J.R., Morton,D.G. and Cheng,N. (1988) Identification of genes required for cytoplasmic localization in early embryos of *C.elegans. Cell,* 52, 311–320.

48. Hyman,A.A. and White,J.G. (1987) Determination of cell division axes in the early embryogenesis of *Caenorhabditis elegans. J. Cell Biol.,* 105, 2123–2135.

49. Laufer,J.S., Bazzicalupo,P. and Wood,W.B. (1980) Segregation of developmental potential in early embryos of *Caenorhabditis elegans. Cell,* 19, 569–577.

50. Schierenberg,E. (1987) Reversal of cellular polarity and early cell–cell interaction in the embryo of *Caenorhabditis elegans. Dev. Biol.,* 122, 452–463.

51. Hill,D.P. and Strome,S. (1988) An analysis of the role of microfilaments in the establishment and maintenance of asymmetry in *Caenorhabditis elegans* zygotes. *Dev. Biol.,* 125, 75–84.

52. Wood,W.B., Schierenberg,E. and Strome,S. (1984) Localization and determination in early embryos of *Caenorhabditis elegans. UCLA Symp. Mol. Cell. Biol.,* 19, 37–49.

53. Schierenberg,E. (1984) Altered cell-division rates after laser-induced cell fusion in nematode embryos. *Dev. Biol.,* 101, 240–245.

54. Schierenberg,E. and Wood,W.B. (1985) Control of cell-cycle timing in early embryos of *Caenorhabditis elegans. Dev. Biol.,* 107, 337–354.

55. Hirsh,D. and Vanderslice,R. (1976) Temperature-sensitive developmental mutants of *Caenorhabditis elegans. Dev. Biol.,* 49, 220–235.

56. Miwa,J., Schierenberg,E., Miwa,S. and von Ehrenstein,G. (1980) Genetics and mode of expression of temperature-sensitive mutations arresting embryonic development in *Caenorhabditis elegans. Dev. Biol.,* 76, 160–174.

57. Cassada,R., Isenghi,E., Culotti,M. and von Ehrenstein,G. (1981) Genetic analysis of temperature sensitive embryogenesis mutations in *Caenorhabditis elegans. Dev. Biol.,* 84, 193–205.

58. Priess,J.R., Schnabel,H. and Schnabel,R. (1987) The *glp-1* locus and cellular interactions in early *C.elegans* embryos. *Cell,* 51, 601–611.

59. Wood,W.B. (1988) Embryology. In *The Nematode Caenorhabditis elegans.* Wood,W.B. (ed.), Cold Spring Harbor Laboratory Press, New York.

60. Horvitz,H.R. and Sulston,J.E. (1980) Isolation and genetic characterization of cell-lineage mutants of the nematode *Caenorhabditis elegans. Genetics,* 96, 435–454.

61. Trent,C., Tsung,N. and Horvitz,H.R. (1983) Egg-laying defective mutants of the nematode *Caenorhabditis elegans. Genetics,* 104, 619–647.

62. Hecht,R.M., Gossett,L.A. and Jeffery,W.R. (1981) Ontogeny of maternal and newly transcribed mRNA analyzed by *in situ* hybridization during development of *Caenorhabditis elegans. Dev. Biol.,* 83, 374–379.

63. Isenghi,E., Cassada,R., Smith,K., Denich,K., Radnia,K. and von Ehrenstein,G. (1983) Maternal effects and temperature-sensitive period of mutations affecting embryogenesis in *Caenorhabditis elegans. Dev. Biol.,* 98, 465–480.

64. Kemphues,K.J., Kusch,M. and Wolf,N. (1988) Maternal-effect lethal mutations on linkage group II of *C.elegans. Genetics,* 120, 977–986.

65. Perrimon,N., Mohler,D., Engstrom,L. and Mahowald,A.P. (1986) X-linked female sterile loci in *Drosophila melanogaster. Genetics,* 113, 695–712.

66. Ito,K. and McGhee,J.D. (1987) Parental DNA strands segregate randomly during embryonic development of *Caenorhabditis elegans. Cell,* 49, 329–336.

67. Strome,S. (1986) Fluorescence visualization of the distribution of microfilaments in gonads and early embryos of the nematode *Caenorhabditis elegans. J. Cell Biol.,* 103, 2241–2252.

68. Strome,S. and Wood,W.B. (1982) Immunofluorescence visualization of germ-line-specific cytoplasmic granules in embryos, larvae and adults of *C.elegans. Proc. Natl. Acad. Sci. USA,* 79, 1558–1562.

69. Wolf,N., Priess,J. and Hirsh,D. (1983) Segregation of germline granules in early

embryos of *Caenorhabditis elegans*: an electron microscopic analysis. *J. Embryol. Exp. Morphol.*, **73**, 297–306.

70. Chitwood,B.G. and Chitwood,M.B. (1974) *Introduction to Nematology*. University Park Press, Baltimore, Maryland.
71. Babu,P. (1974) Biochemical genetics of *Caenorhabditis elegans*. *Mol. Gen. Genet.*, **135**, 29–44.
72. Edgar,L. and McGhee,J.D. (1986) Embryonic expression of a gut-specific esterase in *Caenorhabditis elegans*. *Dev. Biol.*, **114**, 109–118.
73. Cowan,A.E. and McIntosh,J.R. (1985) Mapping the distribution of differentiation potential for intestine, muscle, and hypodermis during early development in *Caenorhabditis elegans*. *Cell*, **41**, 923–932.
74. Schierenberg,E. (1985) Cell determination during early embryogenesis of the nematode *Caenorhabditis elegans*. *Cold Spring Harbor Symp. Quant. Biol.*, **50**, 59–68.
75. Priess,J.R. and Thomson,J.N. (1987) Cellular interactions in early *C.elegans* embryos. *Cell*, **48**, 241–250.
76. Spurway,H. (1948) Genetics and cytology of *Drosophila subobscura*: IV. An extreme delay in gene action causing sterility. *J. Genet.*, **49**, 126–140.
77. Austin,J. and Kimble,J. (1987) *glp-1* is required in the germ line for regulation of the decision between mitosis and meiosis in *Caenorhabditis elegans*. *Cell*, **51**, 589–599.
78. Priess,J. and Hirsh,D. (1986) *Caenorhabditis elegans* morphogenesis: the role of the cytoskeleton in the elongation of the embryo. *Dev. Biol.*, **117**, 156–173.
79. Sulston,J.E. and Horvitz,H.R. (1977) Postembryonic cell lineages of the nematode *Caenorhabditis elegans*. *Dev. Biol.*, **56**, 110–156.

4

Xenopus
Thomas D.Sargent

---

1. Introduction

The study of amphibian embryonic development is an old science; many of the basic concepts and questions were formulated in the 19th century. However, recent advances, particularly in the area of molecular biology, are leading to fundamental changes in the manner in which developmental problems are approached. Molecular markers are complementing, and in some cases replacing, morphological markers as diagnostic characteristics of embryonic events. Inducing factors are being purified and cloned, and the biochemistry of gene regulation is being used to elucidate mechanisms of commitment and determination. The aim of this chapter is to review some of the aspects of *Xenopus* development that have been especially amenable to modern embryology.

This chapter focuses on *Xenopus laevis*, but it should be kept in mind that this frog is by no means the only amphibian whose development has been studied. In fact, much of the classical embryological work was done with other anurans and with various urodeles (salamanders and newts). Although there are important differences, such as the origin of primordial germ cells (see Section 2.1.3), most of the basic developmental processes are conserved between anurans and urodeles. In some instances, work is cited that has made use of species other than *Xenopus laevis* and when this has been done it is so indicated.

1.1 A brief survey of major developmental events

Development begins with oogenesis. As discussed in the next section, the initial processes that determine the organization of the embryo depend on various kinds of information stored in the unfertilized egg, and the manner in which this information is deposited is of considerable interest. Oogenesis in *Xenopus* takes several months to over a year, depending

Table 1.   The stages of *Xenopus* oogenesis according to Dumont (1)

Stage I	Previtellogenic, clear oocyte, diameter up to 300 μm
Stage II	Vitellogenesis, oocyte grows yellow with yolk, diameter up to 450 μm
Stage III	Pigmentation begins, but no obvious animal – vegetal polarity. Diameter up to 600 μm
Stage IV	Animal hemisphere noticeably more pigmented, diameter up to 1000 μm
Stage V	Continued growth to 1200 μm diameter
Stage VI	Full grown oocyte, 1200 – 1400 μm diameter. White band appears at equator.

on environmental conditions. Dumont (1) has published a standard scale for classifying the stages of oocyte development, summarized in *Table 1*. The stage VI oocyte is prepared for fertilization by a process referred to as 'maturation' (see ref. 2 for a review). Maturation is initiated by gonadotropins secreted by the pituitary gland. Follicle cells surrounding the oocyte are stimulated by this signal to secrete the steroid progesterone, which acts at the oocyte surface to trigger its conversion into an egg. This conversion is a complex process; some major events include the disintegration of the giant oocyte nucleus (germinal vesicle), progress of meiosis from the prolonged prophase I to metaphase II, and changes in the cytoskeleton and RNA metabolism. As the mature egg passes through the oviduct it is coated with jelly proteins, which must be present for fertilization to take place (3). In *Xenopus* a single sperm enters the animal hemisphere at a random location (4,5), precipitating a local concentration of pigment which can be readily seen with a dissecting microscope (6). As will be discussed in Section 2.2, the sperm entry point predicts the future ventral side of the embryo. Ninety minutes after fertilization the zygote undergoes first cleavage, which is usually vertical. The following 11 cleavages occur at 25 – 35 min intervals and are approximately synchronous throughout the embryo (7). The result is a 4096-cell blastula with large unpigmented cells at the vegetal and small pigmented cells at the animal pole (*Figure 1A*). A fluid-filled cavity, the blastocoel, is surrounded by a ring of blastomeres known as the marginal zone. At stage 8.5 (4096 cells), a number of dramatic cellular and molecular changes take place, including an abrupt 200-fold increase in the rate of RNA synthesis (8,9), onset of cell motility and desynchronization of subsequent cell divisions (7). Collectively, this set of changes has been referred to as the 'midblastula transition' (10,11). Gastrulation begins shortly after this transition.

Gastrulation is usually considered to begin with the appearance of the blastopore lip on the future dorsal side in the vegetal region (12). However, there are concerted cell movements prior to this, including some migration of presumptive mesoderm around the inner blastopore lip region, and the

A

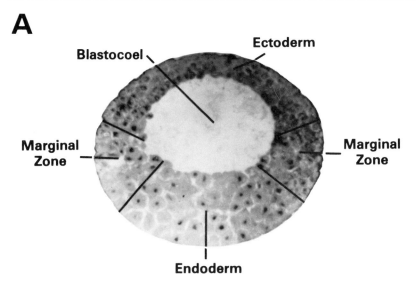

B

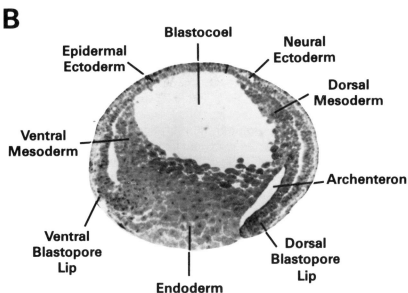

Figure 1. Anatomy of stage 8.5 blastula (**A**) and stage 11 gastrula (**B**) embryos. The approximate diameter of the embryos is 1.2 mm.

beginning of epiboly, or spreading, of the ectodermal surface cells (13). During gastrulation, mesoderm is translocated from the marginal zone upwards along the roof of the blastocoel. This movement is most extensive on the dorsal side where it is driven by 'convergent extension'; dorsal marginal zone cells systematically intercalate, driving a projection into

Table 2.   The stages of *Xenopus* embryogenesis according to Nieuwkoop and Faber (12)[a]

Time (post-fertilization)	Number of cells	Stage
0	1	1
1.5	2	2
2.0	4	3
2.3	8	4
2.8	16	5
3.0	32	6
3.5	(64)	6.5
4.0	(128 – 256)	7
5.0	(1024)	8
5.7	(4096)	(8.5) MBT
7.0	(~7000)	9
9.0	(~10 000)	10
10.0		10.25
11.0		10.5
11.8	(~11 000)	11
13.8	(~12 000)	12

[a]Summarized staging system published by Nieuwkoop and Faber (12) is shown except for stage 8.5, calculated to be the midblastula transition (MBT). These authors do not specify cell numbers after stage 6. Cell numbers through the MBT are shown below, and were calculated by extrapolation from stage 6 onwards, using a cleavage cycle time of 22.5 min, which is the average of cycles 2 through 6. This extrapolation is legitimate since cleavage times are uniform until after the MBT (7). Estimated values are shown in parentheses.

the blastocoel (*Figure 1B*). Lateral and ventral mesoderm do not undergo convergent extension, and the more limited movement of these cells into the blastocoel is dependent primarily upon active migration across the blastocoel roof (14). Gastrulation is considered completed when the mesodermal invasion of the blastocoel ends and the blastopore has constricted and almost closed. Neurulation and organogenesis proceed rapidly, starting with the formation of the neural plate, sensorial placodes, cement gland, somites and notochord. By the end of 24 h of normal development, the embryo consists of around 100 000 cells, has a closed neural tube, partially formed brain, eye, ear and olfactory rudiments, and fully segregated somites and notochord. The nomenclature for stages through gastrulation, as defined by Nieuwkoop and Faber (12) is shown in modified form in *Table 2*.

1.2   Basic concepts and terms

The body plan of the *Xenopus* embryo, like that of all multicellular organisms, is established by a combination of two fundamental processes:

(i)   acquisition of cell identity by inheritance of anisotropically distributed ooplasmic factors;

(ii)   inductive intercellular communication.

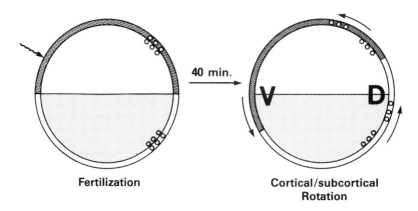

Figure 2. Dorsal–ventral axis determination by cortical–subcortical rotation. The animal hemisphere cortex is shown shaded, although in reality, this region contains very little pigment in *Xenopus*. The stippled area is yolk-laden vegetal cytoplasm. The inner and outer small open circles represent the Nile blue and plasma membrane-bound dye spots used by Vincent *et al.* (15) to visualize cortical rotation, which is counter-clockwise towards the sperm entry point, as indicated. Note that in this drawing the cortex is represented with about ten times its actual thickness relative to the egg diameter. D, dorsal; V, ventral.

Discussion of these processes is facilitated by the use of certain terms, such as 'fate', 'commitment', 'specification', 'determination' and 'determinants'. The use of these words can be confusing if they are not defined explicitly. *Fate* refers to the ultimate destination in the embryonic body of a particular cell or area. In *Xenopus* a fate map is possible because the body plan correlates in an approximately consistent way with the axes of the fertilized egg; ectoderm is derived largely from the animal pole, endoderm from the the vegetal pole, dorsal structures arise on the side opposite the sperm entry point (*Figure 2*, Section 2.2), and so on. The fate map is, however, statistical in nature. There is significant mixing of cells within and between germ layers until gastrulation is completed. There is also some variability between eggs in the position of cleavage planes. Another important point to bear in mind is that the fate map does not imply actual differentiation or bias towards a specific developmental result, it only predicts what is likely to happen if normal embryogenesis takes place.

As the fertilized egg cleaves, the product blastomeres begin to acquire specific identities, by means of one or both of the processes referred to above. This acquisition of identity is referred to as *commitment*, and can be experimentally detected by removing a cell or tissue from its normal position and transferring it somewhere else, either into a neutral environment, that is cultured in saline, or inserted into a different and inductively active region of the embryo. The explant is said to be *specified* (16) if it develops according to its original fate when placed in neutral medium, and *determined* if it does so when exposed to different inductive influences. By definition, a determined cell is also specified, but a specified cell is not necessarily determined. Determination is essentially irreversible while specification implies some degree of plasticity or pluripotency. Examples of this plasticity include the neuralization of dorsal epidermis (Section 3.1.3); lens induction by grafted optic cups in ventral epidermis (Section 3.1.4); and respecification of ectoderm as mesoderm by exposure to inducing factors (Section 3.2).

2. Commitment by cytoplasmic determinants

2.1 The animal – vegetal axis

The *Xenopus* egg has a visually conspicuous polarity along its vertical axis. The upper, or 'animal' hemisphere, is heavily pigmented, and the lower or 'vegetal' hemisphere is not. There is also internal animal – vegetal polarity. The vegetal hemisphere is packed with a dense array of large yolk platelets, while the animal cytoplasm has much smaller platelets and less yolk overall. In the oocyte, the germinal vesicle is located in the animal hemisphere.

A more subtle animal – vegetal polarity is exhibited by the intermediate filament cytoskeleton. Intermediate filament proteins are an exceptionally diverse family of cytoskeletal components comprising type I and type II keratins, vimentin, desmin, neurofilament proteins and glial fibrillary acidic protein (17), and lamins, which are nuclear envelope components (18,19). Godsave *et al.* (20) found that keratin filaments, usually considered characteristic of epithelial cells, were abundant in the oocyte cytoplasm. These filaments were found to be especially concentrated in the cortex, just below, and extensively intertwined with the dense microfilament network associated with the plasma membrane (21). The animal – vegetal polarity of the keratin cytoskeleton has been clearly visualized by using a monoclonal antibody specific for a 56 kd type II keratin, probably corresponding to endo A (22,23). In the vegetal hemisphere, cortical keratin filaments were found to be organized into a highly regular network pattern, while the filaments in the animal hemisphere were more diffuse. This pattern is highly dynamic following maturation. The keratin filaments disappear when germinal vesicle breakdown occurs, and then re-form slowly, leading to a completely cortical pattern at the end of first cell cycle. As shown by Godsave *et al.* (24), there is also a substantial amount of vimentin in the oocyte. Filaments of this protein are distributed throughout the cytoplasm, with concentrations associated with the germinal vesicle and the germ plasm at the vegetal pole (24,25). In contrast to the highly cortical keratin, the vimentin becomes quite evenly distributed after oocyte maturation.

While most tissue differentiation of *Xenopus* depends upon inductive interactions, there are three examples of specification that are probably due to inheritance of egg determinants: epidermis formation from animal cells; primordial germ cell specification by vegetal pole germ plasm; and the transient capacity of vegetal blastomeres to induce animal cells to become mesoderm.

2.1.1 Specification of animal hemisphere cells as epidermis

The third cleavage is usually horizontal and unequal, generating four small animal hemisphere cells and four large vegetal cells. When explanted, the animal blastomeres differentiate spontaneously as somewhat disorganized but recognizable epidermis (26). Vegetal blastomeres develop variably, depending on the amount of marginal cytoplasm present, but usually do not form epidermis and often dissociate spontaneously without differentiating. Therefore, the animal-derived blastomeres and their descendants are specified as epidermis as soon as they are created by division of the fertilized egg. This specification is reversible until the end of gastrulation, and a large fraction of presumptive epidermis is diverted by induction into other pathways, such as the central nervous system.

The mechanism of this epidermal commitment is unknown, but the simplest model postulates a determinant in the animal pole cytoplasm, not necessarily a single substance, that is partitioned to presumptive ectodermal cells during cleavage. Such a determinant would not be tightly associated with the cortex or pigment granules, however. The fertilized egg cytoplasm can be rearranged by inversion and centrifugation (27), which forces the mass of large yolk platelets into the animal hemisphere. The pigment granules remain in place, and when the inverted embryo gastrulates the blastopore appears in the pigmented hemisphere. Such embryos do not survive past late neurula stages, but they clearly form their epidermis from the non-pigmented portion of the egg. Presumably, whatever specifies epidermis can be relocated from the animal to the vegetal pole by centrifugation. Another trait of cells derived from the animal hemisphere is their plasticity and responsiveness to inductive signals. Animal caps, explanted during blastula stage, can be induced to form mesoderm by a variety of growth factor-like signals (see Section 3.2). The ectoderm is also the target of neural induction during the gastrula period. This receptiveness to induction can also be thought of as an autonomously acquired characteristic, complementing the inherent capacity of the vegetal cells to produce mesoderm inducer.

2.1.2 The vegetal pole

According to the fate map, the digestive system arises from the large yolky cells of the vegetal pole (28,29). This fate is apparently not specified until the seventh or eighth cleavage. Miyahara *et al.* (30) showed that dissociation and dispersion of blastulae prior to stage 7 – 8 prevented the later accumulation of alkaline phosphatase, a gut-specific enzyme. When embryos were dispersed after stage 8, alkaline phosphatase accumulated normally, suggesting that gut is at least partially specified by an inductive interaction occurring in midblastula. The endoderm has other roles in *Xenopus*, however, that are probably dependent on vegetally localized information. One is as the source of the germ line. In the egg and early cleavage stages, a distinctive granular material known as germ plasm can be observed near the vegetal pole. This material is partitioned into small cells during mid-cleavage, specifying them as primordial germ cells (PGCs), precursors to the gametes (31,32). UV-irradiation of the vegetal pole results in sterility in the surviving adults. Smith (33) showed that this effect could be reversed in *Rana pipiens* by injecting unirradiated vegetal pole cytoplasm. UV-irradiation of *Xenopus* oocytes, eggs and embryos has a similar effect (34), and there is good evidence that the germ plasm in this species contains material that is able to specify blastomeres as PGCs (35). The PGCs migrate to the genital ridges, completing their specification as germ cells by inductive interactions during this period (36,37). It should be noted that in contrast to these anurans, urodeles generate their PGCs by a mechanism that appears to be entirely inductive (38).

As will be discussed more extensively in Section 3, the vegetal cells also are the apparent source of the signal that induces mesoderm formation. Early vegetal, but not animal, blastomeres produce this inductive signal autonomously when explanted, and this trait is thus likely to be due to inheritance of egg cytoplasmic determinants.

2.1.3 Localized mRNAs in oocytes

Awareness of the animal- and vegetal-specific autonomous developmental fates discussed above has prompted extensive searches for informational molecules that could serve as determinants. For theoretical as well as practical reasons, the search has focused on maternal mRNAs. Preliminary indications that some maternal poly(A)$^+$ RNAs might be distributed unevenly came from the work of Carpenter and Klein (39) who prepared egg animal and vegetal pole poly(A)$^+$ RNA and cDNA, and used this material in solution hybridization experiments. They found that 3–5% of the vegetal poly(A)$^+$ RNA was 2–20-fold less abundant, and possibly absent, in the animal pole. On the other hand, the vast majority of egg mRNA sequences were found to be evenly distributed in both hemispheres. Another approach was taken by King and Barklis (40) who prepared animal and vegetal oocyte RNAs, translated them *in vitro*, and analyzed the products by two-dimensional gel electrophoresis. Five protein species, in the 16–26 kd range and each representing 0.04–0.3% of the mRNA pool, appeared to be highly specific to the vegetal fraction. Rebagliati and co-workers (41) prepared radiolabeled cDNA probes from the extreme animal and vegetal regions of unfertilized eggs and used these to screen an oocyte cDNA library. They identified four different mRNAs that have specific localizations: three, An1, An2 and An3, were largely confined to the animal pole; and one, Vg1, to the vegetal pole. An2 encodes the α subunit of mitochondrial ATPase (42), but the sequences of An1 and An3 have not been reported. Vg1 encodes a polypeptide with 48% homology, in the carboxyl-terminal 114 residues, to the *Drosophila* gene *decapentaplegic* (43,44). The predicted protein products of Vg1 and *decapentaplegic* are members of a family of genes that includes transforming growth factor-β (TGFβ), inhibins, activins and Mullerian inhibitory substance (45,46). As discussed below (Section 3.2) there is fairly convincing evidence that TGFβ-like species are involved in the induction of dorsal, and perhaps lateral and ventral, mesoderm during blastula and early gastrula in *Xenopus*. No other vegetally localized mRNA sequence has been cloned to date and it is possible that Vg1 is unique, at least within the abundance range accessible to the types of screening used.

2.2 The establishment of dorsal – ventral polarity

The unfertilized *Xenopus* egg is radially symmetrical around the animal – vegetal axis. The sperm enters at any point in the animal hemisphere, setting in action a chain of events that establishes the dorsal – ventral axis, usually with the dorsal structures developing on the side opposite the sperm entry point (4,47). Gerhart and colleagues have shown that the critical process that sets up the dorsal – ventral axis is a rotation of the membrane and cortex relative to the deeper cytoplasm, the direction of which is biased, but not strictly determined, by the point of sperm entry (*Figure 2*; 48). The cortex is a layer of submembrane cytoplasm approximately 5 μm thick, devoid of organelles but containing extensive actin and keratin filaments (see Section 2.1.1). The cortex of *Xenopus* eggs is nearly transparent, but in other amphibians, such as the frog *Rana pipiens*, this layer contains pigment granules in the animal hemisphere. The cortical rotation in these pigmented species results in the appearance of a 'grey crescent' at the equator on the dorsal side, as less pigmented vegetal cortex moves up and exposes the underlying endoplasm. The cortical – subcortical rotation has been visualized in *Xenopus* by applying a pattern of dye spots to the plasma membrane and another stain pattern to the subcortical cytoplasm by diffusing Nile blue, which binds to yolk platelets, into the egg through a metal grid (15). After the first cleavage cycle has progressed approximately 40% towards completion, the two sets of spots are observed to begin relative motion, ending up separated by approximately 30° of arc. Coincident with this rotation, the zygote cytoplasm becomes relatively rigid (49,50), due at least in part to repolymerization of microtubules that dissociate shortly after fertilization. The Nile blue stains yolk platelets that are within a few μm of the plasma membrane (51), suggesting that the shear zone separating the cortex and endoplasm is quite close to the egg surface.

If the cortex is immobilized by embedding the zygote in gelatin, the entire endoplasm rotates as a coherent sphere. Dye patterns made in the animal hemisphere shift to the same extent as observed in the vegetal hemisphere, then undergo more complex motion, generally converging towards the sperm entry point, and then fading due to the increasing fluidity of the animal hemisphere cytoplasm. Since this rotation opposes gravity, there must exist some work-generating mechanism which presumably operates in the shear zone.

2.2.1 Experimental perturbation

Grant and Wacaster (52) discovered that UV-irradiation of the vegetal pole of the *Xenopus* zygote resulted in the formation of axis-deficient embryos, which was interpreted as the destruction of a cytoplasmic factor that specified dorsal differentiation (53,54). Recent work suggests a less direct effect. Using the cortical–subcortical dye marking technique, Gerhart and colleagues (reviewed in 47,51) showed that UV-irradiation suppressed cortical rotation. Furthermore, Scharf and Gerhart (55) demonstrated that it was possible to rescue UV-irradiated embryos by orienting them 90° off axis prior to first cleavage. This resulted in virtually normal embryos, with their dorsal sides always forming on the side of the zygote that was facing up. Gravity propels the dense yolk downward, apparently substituting for the rotation blocked by UV. Cleine (56) observed that such gravity-driven rotation did not move the outer $10-20$ μm of subcortical cytoplasm relative to the cortex in the region of the vegetal pole. Furthermore, Scharf *et al.* (57) noted that the cortex of UV-irradiated eggs was particularly difficult to remove and suggested that the UV-irradiation might photo-crosslink the cortical and subcortical cytoplasm, preventing relative movement of these two layers. Taken together, these facts imply that the axis establishment mechanism associated with rotation need not take place precisely at the natural shear zone, which is probably a few μm from the egg surface, and may involve an interaction between the cortex and cytoplasm located further inside the egg.

   Cold, vinblastine, nocodazole and hydrostatic pressure have effects similar to that of UV-irradiation (58). These treatments are known to disrupt microtubules, indicating a requirement for the integrity of these cytoskeletal structures in the rotation mechanism. UV-irradiation might also act by disrupting microtubules, but the UV-sensitive period precedes extensive tubulin polymerization (50). In fact, Holwill *et al.* (34) showed that UV-irradiated oocytes, matured *in vitro*, implanted into females and recovered as fertilizable eggs were axis-deficient, so whatever was damaged by UV could not be repaired or replaced during the 40 h that elapsed between irradiation and fertilization.

   Another treatment which affects dorsal–ventral axis determination is exposure of the fertilized egg, during the period of 0.2 to 0.35 of the first cycle, to heavy water, D_2O (48,59). This has the effect of hyper-sensitizing the zygote to rotation; even a few degrees of displacement can result in normal axis development. Some rotation must occur, however; even D_2O-treated embryos will be axis-deficient if there is no displacement. More extensive rotation results in embryos with exaggerated dorsal and anterior phenotypes, the extreme version of which is an oversized, radially symmetrical head, with circumferential cement gland and eye tissue. Most of the marginal zone in these embryos develops into heart and head mesoderm. Thus graded exposure to rotation-suppressing or response-enhancing treatments can lead to a spectrum of axial abnormalities, ranging from completely ventral – posterior to completely dorsal – anterior. While the extent of rotation roughly correlates with the degree of the dorsal – anterior phenotype, this movement should be considered necessary but not sufficient for axial specification, since in some cases axis-deficient embryos develop even when cortical – subcortical rotation appears to be normal (48).

3. Induction

Cell identity that is not specified by inheriting determinants from the egg cytoplasm must arise via inductive interactions. *Xenopus* embryos have been instrumental in the experimental identification and analysis of such interactions, due to their large size, relative ease of manipulation and the existence of a roughly consistent fate map, which makes it possible to discern changes in commitment that result from exposing embryonic cells and tissues to suspected inductive signals. These experiments can be done by transplanting competent tissue to an ectopic position, exposing competent tissue to extracts containing inducer (e.g. 60), or by dissociating embryos into individual cells, which prevents intercellular communication (30,61). If any of these manipulations alters the developmental fate of the test tissue, then an inductive interaction is implicated. This section discusses several examples of inductions identified in *Xenopus* embryos, from early cleavage through neurulation.

3.1 Patterns of induction

3.1.1 Mesoderm induction
One of the most influential observations in modern amphibian developmental biology derived from the animal – vegetal recombinations carried out by Nieuwkoop and colleagues (62). Explants from the extreme animal and vegetal regions were removed from midblastula embryos, discarding the marginal zone. The explants were cultured either separately or combined. In the former case, the animal cap differentiated into epidermis, its autonomous pathway, while the vegetal core failed to

differentiate detectably. When combined, however, a large amount of mesoderm was generated, arising from cells that would have otherwise differentiated as ectoderm. The recombinants formed the full range of mesoderm, ventral, lateral and dorsal, and proceeded to carry out neural and other secondary inductions.

Two things happen during this inductive interaction. The animal hemisphere cells, destined to become ectoderm, are diverted into the mesodermal pathway, and at the same time, the dorsal–ventral polarity of the vegetal core is imposed on the transplanted ectoderm both with axolotls (63), and with *Xenopus* (64). Both of these effects result from induction of ectoderm by endoderm, but it is not clear how they are related mechanistically. The simplest model (47) proposes that an inducing factor is secreted by vegetal cells of the dorsal quadrant generating a dorsal–ventral concentration gradient, which results in dorsal specification at one end of the marginal zone, and lateral and ventral specification, respectively, in the middle and ventral areas. Circumstantial support for such a model comes from the observation of such a graded dorsal–ventral response of animal hemisphere explants to mesodermal inducer purified from the *Xenopus* tissue culture cell line, XTC (Section 3.2; 65). On the other hand, Slack and colleagues (64,66,67) have argued for distinct signals that independently specify ventral and dorsal mesoderm. Whichever model proves correct, it is clear that mesoderm induction is a complex phenomenon that results in the formation of a diverse range of cell types. Furthermore, most of the commonly used markers for 'mesoderm' are muscle-specific proteins such as α-actin or myosin, and thus the expression of these markers is the result of *dorsalization* in addition to, or in some experimental contexts, instead of, *mesoderm induction*. The remainder of this section deals with general properties of mesoderm induction, and interactions that are partly or entirely specific for dorsalization are discussed in the next section.

The period in which mesoderm induction can be elicited in animal–vegetal recombinants is limited. Heterochronic grafting revealed that the vegetal cells can produce inducing signals by stage 5, and possibly earlier (68). The inducing capacity of these cells increased gradually through cleavage, and was lost in the extreme vegetal region by stage 9 (63,69,70). Animal cells acquired the ability to respond by stage 6.5, and lost competence to respond by stage 10.5 or shortly thereafter (68,69). Since the minimum contact time between hemispheres for (dorsal) mesoderm specification is 1.5–2.5 h (69), this induction should be complete by the time of the midblastula transition (stage 8.5, see *Table 2*).

One point that is important to bear in mind is that these experiments only demonstrate the *capability* of vegetal cells to induce animal hemisphere cells to form mesoderm rather than ectoderm, their normal fate. The notion that an equivalent inductive process takes place in normal embryonic development is almost certainly correct, but has not been

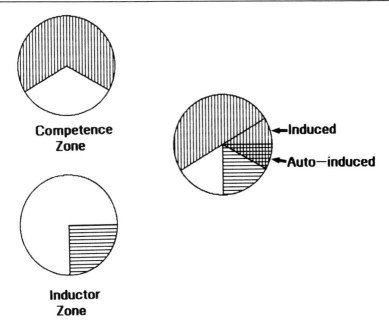

Figure 3. Schematic diagram for autonomous and induced specification of dorsal mesoderm. The competence zone (shaded) refers to the region which is capable of responding to vegetal inductive signals, and the inductor zone refers to the region that generates the dorsal- and mesodermal-inducing signal(s). The overlapping area is capable of self-induction, and the area directly above can be induced by underlying cells to enter the dorsal mesoderm pathway.

proven. If the formation of mesoderm were due entirely to interactions between the extreme vegetal and animal pole regions, and if the time constraints were similar to those of recombinants (see above) then marginal zone explants taken prior to stage 8.5 would not yet be specified as mesoderm. In fact, groups of cells explanted from the marginal zone early in cleavage will subsequently differentiate mesoderm, suggesting that the marginal zone has some autonomous mesodermal identity (71,72). The only evidence that induction is required for mesodermal specification of the marginal zone was the observation (61) that dissociating and dispersing embryos throughout the cleavage period blocked the accumulation of muscle-specific mRNA. Strictly speaking, this is evidence only for obligatory inductive interactions in muscle or dorsal–lateral mesoderm development. The status of non-myotomal mesodermal commitment under such conditions is not known. One way to accommodate both the marginal zone explant and dispersion data is to postulate that during normal development a portion of the marginal zone autonomously acquires the ability to induce itself via a mechanism that depends on close cell proximity. Such a model is schematically outlined in *Figure 3*. Receptors for inducing signals could be located in the animal hemisphere, dorsal mesoderm-inducing capacity located in the dorsal half of the vegetal hemisphere, and a region of overlap in the lower portion of the dorsal marginal zone. Cells in the overlap region could both make inducer and respond to it, and would therefore exhibit autonomous dorsal mesodermal character that nonetheless depends upon providing an environment, such as close cell contact, that permits the secreted inducer to interact with receptor.

3.1.2 Dorsalization

A one-day-old *Xenopus* neurula has morphologically distinct dorsal – ventral and anterior – posterior axes. Examination of disrupted embryos reveals that these axes are in fact closely related. For example, the cement gland is located on the ventral side of the head, close to where the mouth will form. However, this structure is among the most sensitive to the effects of preventing the cortical – subcortical rotation prior to first cleavage (see Section 2.2) suggesting that its formation is part of the dorsal program of development that includes formation of muscle, notochord and neural tube. Similar conclusions can be drawn from examining the results of treating blastulae with lithium (73), bisecting dispermic zygotes (74), or ablating portions of the early cleavage embryo (75,76,77). Thus dorsal specification leads to formation of both dorsal and anterior structures, some of which are located ventrally in the larval body. This section briefly reviews several sets of experiments that reveal the existence of interactions that can dorsalize competent tissues.

The earliest event, following the cortical – subcortical rotation, that involves inductive dorsal axis specification may take place at the eight-cell stage. Cardellini (78) removed the four small animal blastomeres at this stage, rotated this explant 180° and placed it in contact with four large vegetal cells from another embryo. In most cases in which the embryo survived, the dorsal axis corresponded to that of the animal cells. One cell division later, at the 16-cell stage, the same experiment yielded the opposite result; the vegetal cells, now eight in number, determined the dorsoventral axis. As is always the case with this type of experiment, there is no straightforward way to prove that in normal development the four animal micromeres actually impose an axis on the vegetal cells. They appear to have that capability, however, at least for a brief period. By the 32-cell stage it is clear that cells of the vegetal hemisphere can determine the dorsal axis of the rest of the embryo. This has been shown in two ways. Dale and Slack (64) combined cells from the dorsal, lateral or ventral quadrants of the vegetal-most tier with the animal-most tier at the 32-cell stage and noted the type of tissues forming from the animal tier cells. The dorsal vegetal blastomeres showed a strong tendency to induce dorsal mesodermal derivatives such as notochord and muscle, while the lateral and ventral blastomeres induced mainly ventral mesoderm, such as blood, mesenchyme and mesothelium. A similar result had been obtained with axolotls by Boterenbrood and Nieuwkoop (63), who also made the interesting observation that the dorsalizing inductive potential of dorsal vegetal blastomeres was lost in an all-or-nothing manner, rather than declining through a transition state in which they induced lateral or ventral mesoderm. In work with the axolotl, dorsal blastomeres were also found to lose inducing activity sooner than ventral blastomeres. The second demonstration comes from the work of Gimlich and Gerhart (70,79). These experiments are based on the dorsal suppression that results from UV-irradiation (Section 2.2). Such embryos lack all dorsal and anterior structures, but are unimpaired in their ability to form ventral mesoderm. Normal or approximately normal embryogenesis can be restored by transplanting pairs of vegetal dorsal blastomeres from non-irradiated donors into irradiated hosts. These same cells, when transplanted into the ventral region of non-irradiated hosts, frequently lead to the formation of a second dorsal axis, yielding 'Siamese twins'. Ventral cells, or any cells from the animal region, had no rescue effect when transplanted. When the donor cells were derived from the lowest vegetal tier, axis restoration was achieved entirely by induction, that is the donor cells did not contribute directly to any of the axial structures. Gimlich (70) found that at the 32-cell stage, the next to the lowest tier (tier 3) also possessed dorsalizing capability, although at a level less than that found at later stages. Cells from this region also participate in forming axial structures, such as muscle and notochord, that is they both develop into and induce dorsal mesoderm. It is apparent from experiments in which tier 3 is explanted as a block of tissue that this tier is in fact specified as dorsal mesoderm by the 64-cell stage, and perhaps earlier (70,71). As mentioned above, this specification depends at least partly on intercellular interactions amongst daughter cells arising within the explant, and the autonomy may be due to a self-induction as described in *Figure 3*.

Slack and colleagues (64,80) have presented evidence for a dorsalization event occurring in the marginal zone late in the blastula stage. Until this point in development, mesoderm from the lateral regions of the marginal zone is specified, as judged by explantation experiments, as extreme ventral, differentiating into mesothelium, mesenchyme and blood. Combining ventral or lateral with dorsal marginal zone explants results in respecification of the lateral cells to a more dorsal or intermediate level, leading to these three tissue types plus some muscle. The dorsal cells do not become more ventral in such combinations.

It is worth briefly reconsidering at this juncture the findings of Curtis (81,82) that grey crescent (i.e. future dorsal, see *Figure 3)* cortex could induce a second axis when transplanted into the future ventral area of the zygote. Gerhart *et al.* (5) argued that Curtis' results could be explained by the gravity-driven rearrangement of vegetal cytoplasm during microsurgery (which involved holding recipient zygotes in an off-axis orientation), thus generating twins. This argument weakens but does not disprove Curtis' conclusions that the grey crescent cortex has dorsal inductive potency. In fact, Tomkins and Rodman (83) transplanted this cortex into the blastocoel of midblastula *Xenopus*, and found significant induction of second axes. The frequency increased when the cortex was removed from four-cell embryos, and there was some induction with animal pole cortex, but neither of these results is inconsistent with the other findings discussed above. The possibility must be kept open that the dorsal cortex or closely associated cytoplasm, contains some activity which is able to induce a second dorsal axis in recipients as late as midblastula.

3.1.3 Neural induction

All vertebrates form their CNS in a generally similar manner; a portion of the ectoderm is diverted from epidermis and induced to form a neural plate by an interaction with dorsal mesoderm. These two tissues are brought into close contact by gastrulation movements. The neural plate is deformed into a tube, which subsequently differentiates along the cranial – caudal axis into brain and spinal cord. The interactions responsible for the various developmental decisions are likely to be numerous and spread out over the entire period of organogenesis.

The history of investigation of amphibian neural induction has been afflicted with a certain amount of confusion and misconception. One persistent problem has been the erroneous designation of neural induction as 'primary', which obscures the important and earlier roles of mesoderm induction and dorsalization (64). Another problem stems from the extent to which gastrula ectoderm, particularly that of urodeles, is predisposed to switch from an epidermal to a neural pathway. This property complicates the identification of legitimate neural inducing substances. *Xenopus* appears less prone to artifactual neural induction, but nevertheless, experiments involving putative neural inducers have to be interpreted very carefully. Finally, the CNS is a complex organ, and its differentiation certainly involves multiple inductive interactions which may be rather difficult to distinguish from one another.

One fact that has been established beyond doubt is that induction is required for neural plate formation. Pre-neural ectoderm removed from the embryo prior to gastrulation does not form a neural tube, nor does this explanted tissue express neural markers such as neural cell adhesion molecule (NCAM; 84), neurofilament protein (85), or XlHBox6, a neural-specific homeobox-containing sequence (86). The ectoderm – mesoderm contact during gastrulation results in suppression of epidermal markers such as keratin and activation of NCAM (84,87,88). These responses can be detected within an hour after ectoderm contacts dorsal mesoderm, which implies that neural induction, at least the initial steps, occurs more rapidly than mesoderm induction, which requires about 1.5 – 2.5 h (69). There is probably a connection between the rapidity of initial neural induction and the neural predisposition of ectoderm.

This predisposition also appears to correlate with the dorsal – ventral polarity of the gastrula. Sharpe *et al.* (86) showed that dorsal sectors of the early gastrula ectoderm were more readily induced to express neural markers (NCAM and XlHBox6) than were ventral sectors when combined with dorsal mesoderm, suggesting that the animal hemisphere is partially imprinted with dorsal – ventral identity prior to gastrulation. It is conceivable that this specification takes place as early as during the initial establishment of the dorsal – ventral axis, that is the cortical rotation of the first cell cycle.

As noted above, it is clear that neural development depends upon induction during gastrulation of dorsal ectoderm by underlying mesoderm. While this process can be thought of as part of a general dorsalization program, it is not clear how it is related to other components of dorsalization, such as muscle and notochord formation. We have recently completed experiments bearing on the question of how the chordamesoderm acquires the capacity to induce neural tissue. Induction during blastula to early gastrula stages is required for the activation of muscle actin (61). Since muscle is such a large component of dorsal mesoderm, it might be predicted that the dependence of muscle specification on induction could be extrapolated to dorsal mesoderm in general, including the capacity of this tissue to induce CNS. This prediction is not correct. Embryos dissociated and dispersed from stage 7 through stage 10.5, when reaggregated, failed to accumulate any muscle-specific mRNAs, but accumulated NCAM mRNA to the same extent as controls. This neuralization is not an artifact of dispersion. When dissected animal caps, known to be competent to respond to neural induction, were dispersed and reaggregated under conditions similar to those used for whole embryos, they did not accumulate NCAM mRNA. One interpretation of these results is that the inductive dorsalizing capacity of dorsal vegetal cells is maintained through cleavage in a cell-autonomous fashion and is still present and active in gastrulae, at which time it can complete the dorsalization of juxtaposed competent ectoderm, triggering its neural specification. Muscle would not be induced in such a situation as the competence period for this inductive interaction terminates by stage 10.5 (69).

3.1.4 *Inductive organogenesis*

Following the specification of epidermis, discussed in Section 2.1.1, the earliest examples of organogenesis can be found in the induction of anterior ectoderm to form the cement gland and the sensorial structures of the head. These tissues are specified by a sequence of inductive interactions between the head ectoderm and the tissue that underlie it (89). During gastrulation the pharyngeal endoderm and presumptive heart mesoderm, leading the involuting chordamesoderm, first make contact with anterior ectoderm, resulting in the establishment of the lens, olfactory and otic placodes. Subsequent contacts with regions of the brain elicit and organize the actual formation of the lens, nose and ear (89). The multistep nature of these inductions has been emphasized recently by work of Grainger and colleagues, who have studied the interactions involved in lens formation (90,91). Classical work of Spemann and others had suggested that the lens was induced in ectoderm exclusively via an interaction with the optic cup region of the brain. The primary result leading to this conclusion was that optic cups transplanted to ventral ectoderm of mid to late neurula elicited lens formation. Using lineage-marked grafts, Grainger and colleagues have shown that this ectopic lens formation is not induced in overlying ectoderm but is rather the result of delamination of lens cells from the transplanted cup. However, ectoderm could be induced to form lens at the midgastrula stage, at which time the future head ectoderm was 2–3 times more readily induced than ventral ectoderm. By stage 15.5, the only ectoderm capable of being efficiently induced to form lens was in the head region, a competence that is probably specified by the above mentioned interaction with heart mesoderm and pharyngeal endoderm during mid to late gastrula.

3.2 Polypeptide growth factors as inducers

A recent advance in the study of amphibian induction has been the discovery that the inducing signals for general mesoderm induction and for dorsalization are related to mammalian polypeptide growth factors. Both acidic and basic fibroblast growth factors (FGF), as well as other heparin-binding growth factors, are able to induce explanted ectoderm to become ventral, lateral and to some extent dorsal mesoderm (66,92). Slack and colleagues (66) have shown that vegetal cells can respecify ectoderm as mesoderm in explants separated by a nylon mesh, and that this induction is inhibited by addition to the medium of high concentrations of heparin. FGF binds heparin avidly, and the observed inhibition could have been due to heparin competing with ectodermal receptors for FGF, or other heparin-binding growth factors emanating from the vegetal cells. Another result suggesting a role for FGF was the identification of a mRNA in the *Xenopus* oocyte encoding a protein nearly identical to bovine basic FGF (93). It has also been reported that while FGF alone induces low levels of α-actin in ectodermal explants, combining TGFβ and FGF significantly enhances this effect (93). Rosa *et al.* (94) found that TGFβ2, but not TGFβ1, was much more active than FGF in inducing α-actin in animal caps. Some response could be detected with as little as 3–12 ng ml^{-1} of TGFβ2, while relatively high concentrations (200–400 ng ml^{-1}) induced α-actin levels similar to that obtained with animal–vegetal recombinants. Rosa and co-workers also observed some relatively weak synergistic interaction of TGFβ2 and FGF (94). There are several possible sources of variability in experiments of this kind. One hazard is contamination of the growth factor preparations, a serious issue as these factors are often biologically active at concentrations 10 – 100 times lower than those used in these induction experiments. Another problem is variations in the details of the experimental designs; the time of exposure to inducer, the exact concentrations used, the size of the explants, and other factors may have unexpected effects on the outcome, so findings from different laboratories need to be evaluated carefully.

A less characterized but extremely potent mesoderm-inducing factor has been detected and purified to apparent homogeneity from medium conditioned by the *Xenopus* tissue culture cell line, XTC (60,65). This XTC factor induces ventral mesoderm at relatively low concentrations ($0.2 – 1$ ng ml^{-1}), muscle at intermediate concentrations ($1 – 5$ ng ml^{-1}) and notochord at high concentrations ($5 – 10$ ng ml^{-1}). This graded response demonstrates that in principle, a single factor could be responsible for the entire range of mesodermal differentiation *in vivo*. Rosa *et al.* (94) showed that a neutralizing antibody specific for TGFβ2 blocked the muscle-inducing capability of diluted XTC-conditioned medium, suggesting that the active ingredient in crude XTC factor is related to TGFβ, although much more active as a mesoderm inducer. There is as yet no evidence that XTC factor is actually present in the embryo, although this should be settled in the near future. Given the similarity of XTC factor to TGFβ, and the presence in the vegetal hemisphere of Vg1 mRNA, which encodes a TGFβ-like polypeptide, it seems probable that one or more members of this family of growth factors will prove to be important in mesoderm induction. An important aspect of these observations is that a wide variety of factors can have rather similar effects on embryonic ectoderm. Low concentrations of XTC factor, medium concentrations of TGFβ2 and high concentrations of either acidic or basic FGF all induce α-actin gene expression. It is unlikely that all these ligands bind to the same embryonic surface receptor, so that similarity of their responses suggests that the mechanisms of their effects merge at some point, such as at the level of signal transduction or modulation of regulatory transcription factors.

A possible clue regarding the mechanism of transduction of the inducing signal comes from observations of the effects of treating embryos with lithium chloride. This has the effect upon stage 6 embryos of giving the ventral and lateral marginal zones the capacity to induce dorsal axial development in surrounding tissue (95). Slack and his colleagues (67) reported that lithium chloride treatment enhanced the dorsal mesoderm induction effects of FGF, an interaction that could be analogous to that of TGFβ and FGF. These results raise the possibility that the cellular response to mesoderm and dorsalizing induction is mediated via the inositol phosphate signal transduction mechanism, which can be disrupted by lithium ions (96). However, lithium also can in some cases directly affect the function of both stimulatory and inhibitory G proteins (97), which are involved in many signal transduction pathways (for a review see ref. 98), so the interpretation of the effect of lithium on embryos is difficult to extend beyond the level of phenomenology.

Investigating the biological properties of polypeptide growth factors has been a major area of research for many years, and much has been learned about how these factors influence the behavior of various cells and tissues. Of particular interest, given their apparent roles in amphibian development, is the ability of TGFβs and, to a lesser extent, FGFs to block the final steps in the terminal differentiation of various cell lines such as pre-adipocytes (99) and myoblasts (100,101). When TGFβ, FGF or serum are withdrawn from these progenitor cells, differentiation gradually is initiated. This process is reversible up until the time when the final tissue-specific gene expression begins, after which point restoration of the growth factors does not halt differentiation, a property that is perhaps analogous to the loss of competence of embryonic cells to respond to inducers. Another observation that may be relevant to mechanisms of embryonic response to induction is that of Olson *et al.* (102), who reported that the inhibitory effects of serum, FGF or TGFβ on myoblast differentiation could be circumvented by transfection of these cells with an activated form of the oncogene *ras*. This oncogene has been implicated as being important in the inositol phosphate signal transduction system, probably downstream from the activity of phospholipase C (103). There is always a degree of uncertainty associated with applying results obtained with cultured cells to intact embryos. On the other hand, the basic mechanisms of intercellular signalling and the manner in which this affects gene regulation are likely to bear some similarity amongst diverse systems, and the biochemistry of growth factors and signal transduction have thus become important areas of developmental biology.

4. Gene regulation

Because of the ease with which *Xenopus* embryos can be obtained in quantity, their large size, their rapid development *in vitro* and the moderate genome size of the organism (similar to mammalian genomes), it has been a fairly simple matter to identify and clone a variety of genes that are regulated during development. The molecular details underlying such gene regulation are important to developmental biology because, in principle, they should provide a starting point for unravelling the mechanistic complexities of embryological phenomena, such as inductive interactions.

This section discusses some of the progress that has been made in the analysis of gene regulation in *Xenopus*. I will first discuss the 5S RNA genes, which are comparatively well understood. The second part of this section will evaluate experiments in which genes have been introduced into *Xenopus* embryos. It will focus upon some recent work with two tissue-specific class II genes encoding muscle actin and epidermal keratin. Finally we will examine some examples of post-transcriptional gene regulation.

4.1 The 5S RNA genes

One of the structural components of ribosomes is 5S RNA, which, in eukaryotes, is transcribed by RNA polymerase III from repetitive genes. In *Xenopus* there are two main families of 5S RNA genes; the major oocyte family, with 40 000 copies, and the somatic family, with 800 copies per diploid nucleus. These genes occur as blocks of tandem repeats, separated by spacer DNA sequences that are quite different between oocyte and somatic families. The oocyte and somatic genes themselves are very similar, with only six differences out of a total length of 120 nucleotides (104–106). The correct initiation of transcription of these genes is controlled by a 50 bp region located internally from +47 to +96, which has therefore been called the internal control region (ICR; 107,108). The ICR can be subdivided into three segments, box A, which is also present in the related ICR of tRNA genes (109), box C and an intermediate sequence lying between these two elements (110) (*Figure 4*). While the ICR specifies the correct initiation of transcription, sequences outside of this region have significance in the relative efficiency of somatic and oocyte templates (111,112).

Transcription of 5S RNA genes requires the assembly of a DNA – protein complex comprising the ICR and at least three transcription

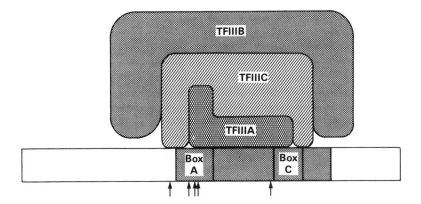

Figure 4. The 5S RNA gene. The 120 nucleotide gene is shown 5′ end to the left, with the ICR stippled. Boxes A and C are domains within the ICR (see text). Arrows point to nucleotide differences between the major oocyte and the somatic 5S RNAs. Note that TFIIIB and TFIIIC are probably multiple polypeptides.

factors, TFIIIA, TFIIIB and TFIIIC (*Figure 4*). TFIIIA is in some ways the most thoroughly characterized of all eukaryotic transcription factors. This protein has dual functions; in addition to its role in class III gene activity, TFIIIA is used to store the enormous amount of maternal 5S RNA in the oocyte as a 7S ribonucleoprotein particle (113–115). This particle can easily be isolated, and TFIIIA can be readily purified to homogeneity, greatly facilitating the study of its interactions with DNA. Full length TFIIIA cDNA clones have been obtained, from which the primary sequence was determined (116). The most striking feature of the TFIIIA sequence is the presence of nine repeating motifs called 'zinc fingers' which are thought to be critical in the interaction of this and many other transcription factors with their binding sites in DNA (117). The other two 5S RNA gene cofactors, TFIIIB and C, are less well characterized. Each probably comprises more than a single polypeptide. The initial interaction in the formation of the transcription complex is between the ICR and TFIIIA. TFIIIC binds subsequently, and is important in stabilizing the complex (see below), probably both by effects on TFIIIA and direct contacts with DNA (109,118). Initiation of transcription requires the participation of all three factors (119,120) in a complex which is extremely resistant to disruption by the passage of RNA polymerase (121,122) but is completely dissociated by transit of a DNA replication fork (123).

In the germinal vesicle, conditions are relatively permissive with respect to 5S RNA gene transcription, and both types of RNA are synthesized at levels approximating their proportions in the *Xenopus* genome. This ratio changes dramatically during early development, and by the late gastrula stage, the oocyte family is repressed and somatic 5S RNA predominates, representing a preference of about 1000-fold for expression of the somatic versus oocyte genes. This transition can be regarded as a model for the general phenomenon of stable changes in gene regulation in development (124). There are two levels to this control; first, the selective inactivation of the oocyte genes, and second, the perpetuation of the inactive state.

The first control level depends on the substantially greater stability of the somatic gene versus the oocyte gene transcription complex. Ribosomes compete with 5S RNA genes for binding of transcription factors (125). This competition leads to the selective inactivation of oocyte genes when the transcriptional apparatus is exposed to the cytoplasm following germinal vesicle breakdown and at subsequent cell divisions. Furthermore, as the genome replicates during cleavage, there is a decline in the ratio of TFIIIA molecules to 5S RNA genes, and this also contributes to loss of oocyte gene expression. The TFIIIA concentration can be transiently elevated during cleavage by injecting synthetic TFIIIA mRNA into fertilized eggs, which results in prolonged expression of oocyte genes (126). However, Peck *et al.* (111) showed that whole oocyte extracts, which express somatic genes 100 times more efficiently than germinal vesicle extracts, generated TFIIIA footprints on both types of gene, suggesting that the disruption of oocyte complexes could occur via removal or modification of other factors, such as TFIIIC (118), without dissociating TFIIIA. Indirect but significant evidence for the importance of TFIIIB and/or TFIIIC comes from the observation that oocyte and somatic ICRs have identical affinity for TFIIIA (127).

The second facet of 5S RNA gene regulation is the stability of the selective inactivation of the oocyte genes. If chromatin is isolated from somatic cells and transcribed *in vitro* with added RNA polymerase III, authentic 5S RNA is synthesized, but only of the somatic family (128). This remains true even if the reaction is supplemented with excess transcription factors, indicating that the oocyte genes have been sequestered in some way. Histone H1 probably plays a major role in this sequestration, as its removal from chromatin by salt treatment permits the programming of oocyte gene transcription by added factors and RNA polymerase III (128). This histone may act by facilitating the condensation of inactive chromatin into higher order structures that are not accessible to transcription factors. Another potential mechanism for oocyte gene repression *in vivo* is the tendency of these genes to replicate later in S phase than the somatic genes (129). Since DNA synthesis disrupts complexes of both families, each round of replication provides an opportunity for competitive recruitment of transcription factors (123). By replicating earlier, the somatic genes could be reprogrammed for transcription while the more numerous oocyte genes were still condensed and inaccessible. This late replication may be the result, as opposed to the cause, of reduced transcriptional activity, but in either case it could be quite important in maintaining the status quo of differential gene expression.

4.2 Gene transfer into *Xenopus* embryos

The *Xenopus* egg is quite large, making microinjection of nucleic acids and other materials a relatively simple procedure. High molecular weight DNA injected after fertilization, often, but not always, replicates during cleavage, and usually disappears gradually during organogenesis. A small amount of injected DNA integrates into the genome (130), but this has not been useful in this system. In spite of the transient presence of introduced DNA, cloned copies of genes that are normally expressed during early embryogenesis can be brought under correct regulation, provided a rapid assay for regulatory elements.

The rapid cleavage of *Xenopus* zygotes generates approximately 10 000 copies of the genome in the few hours before gastrulation. Sufficient protein constituents of chromatin must be stored in the egg to make possible this rapid assembly of nuclei. Laskey *et al.* (131) showed that extracts from unfertilized *Xenopus* eggs could spontaneously package exogenously supplied DNA (SV40) into chromatin indistinguishable from native minichromosomes. A single egg contains enough histones and other proteins to assemble 40 ng of DNA, equivalent to approximately 7000 *Xenopus* nuclei. Purified bacteriophage DNA microinjected into *Xenopus* eggs is spontaneously packaged into bodies that closely resemble nuclei, exhibiting bilayered envelopes, peripheral lamin networks and nuclear pores (132). Furthermore, DNA injected into fertilized eggs usually replicates, often increasing many-fold, during cleavage. This replication activity can also be recovered in egg extracts (133–135).

Thus it is possible to introduce DNA into the embryo with the expectation that it will be properly assembled with chromatin proteins, and perhaps fall under appropriate regulatory constraints. Nevertheless, the initial attempts to realize correct developmental control of injected genes were unsuccessful. Cloned globin genes, for example, normally expressed in late embryogenesis or in adult animals were not regulated properly when introduced into early embryos (136).

4.2.1 The GS17 gene

An important advance came from the work of Krieg and Melton with a gene of unknown function, named GS17, that is normally expressed only for a very brief period during gastrulation. Genomic GS17 DNA, injected as a supercoiled plasmid, was found to be expressed at the correct time (137). Furthermore, a small sequence element from the 5' flanking region was capable of conferring this temporal regulation upon a heterologous promoter from the *Xenopus* β-globin gene (138). The upstream element from the GS17 gene thus represents an enhancer-like element that interacts with a signal present from the midblastula transition until mid to late gastrula. In view of the developmental importance of regulated gene expression during this formative period, the *trans*-acting factors that interact with this enhancer will be of considerable interest.

4.2.2 The muscle-specific actin gene

Subsequent to the work with GS17, two groups (139,140) reported similar results using genes encoding α-actin, which, as noted earlier, is expressed only in muscle. Wilson and colleagues cloned an α-actin gene from the closely related species *Xenopus borealis*, and injected this gene intact, as well as fusions of the 5' region of the α-actin gene to the 3' region of a mouse β-globin gene. The fusion gene contains α-actin sequences from approximately −3000 to a site just downstream from the end of the second exon, and was usually found to be expressed primarily in the myotomes. One experiment was described in which most of the injected, amplified DNA became localized to the ectoderm, resulting in apparent ectoderm-specific expression of the fused gene. This illustrates one shortcoming of such experiments that could lead to misleading results; the injected DNA is not necessarily uniformly distributed throughout the embryo, and adventitious concentrations of template may result in transcript accumulation in inappropriate tissues.

Mohun and colleagues made use of constructs fusing the 5' flanking region of the α-actin gene, ending at $+23$, to a bacterial reporter gene encoding chloramphenicol acetyl transferase which can be detected at the protein or RNA level by sensitive assays. The region from -417 to $+23$ conferred tissue specificity, although constructs with longer 5' regions tended to be expressed at higher levels. In addition, a microinjected cytoskeletal actin gene also exhibited enhanced activity in myotomes. The endogenous copy of this cytoskeletal actin gene is not muscle-specific, so it is possible that the regulation of injected actin genes may not be accurate in all respects. However, it is clear that the expression of this introduced muscle actin gene is at least approximately correct, and furthermore that the DNA sequences involved in mediating this control are located in a few hundred nucleotides of the 5' flanking region. One interesting feature of this region is the presence of several copies of the serum response element (SRE; 141), a control element first described in the context of the *c-fos* gene, which is rapidly induced when cells are exposed to serum growth factors (142). The presence of SREs in the α-actin gene control region is especially intriguing in view of the likely involvement of growth factor-like hormones in the induction of muscle in *Xenopus*. On the other hand, SREs are associated with all actin genes, including non-muscle isoforms, so additional control elements appear to be involved.

4.2.3 Epidermal keratin genes

Our group at NIH has been studying the expression of the tissue-specific genes encoding epidermal keratins. As discussed in Section 2.1.1, the epidermis differentiates autonomously, that is without need for inductive interactions. Among the earliest molecular manifestations of this differentiation is the activation of a battery of genes encoding type I and type II keratins, which make up the characteristic intermediate filament cytoskeleton of epidermal cells. Another important feature of keratin gene regulation is their inactivation by inductive interactions that divert cells from epidermis to mesoderm (61,143) or central nervous system (84,88). Thus keratin genes are controlled both by a positive mechanism, perhaps involving interaction with potential ooplasmic determinants, and by negative interaction with the machinery that responds to induction.

We have found that a cloned keratin gene, specifically the one designated XK81A1 (144), is correctly regulated following injection into the fertilized egg. Experiments with different deletion and fusion constructs reveals that the region from -487 to $+26$ contains all the information necessary for epidermal specificity. Deletions into this region result in loss of promoter activity, possibly accompanied by a relatively minor increase in the activity of the associated promoter in non-epidermal tissues. The control mechanism specifying epidermal expression is thus primarily positive in nature, and work is underway to identify and characterize the protein factors that interact with this region.

4.3 Post-transcriptional gene regulation

Xenopus oocytes have accumulated steady-state concentrations of most poly(A)$^+$ RNA sequences by stage II (145,146), and subsequent synthesis of these RNAs is balanced by an equal amount of decay. The translational apparatus increases in capacity until around stage IV, then remains approximately constant thereafter. During the early oocyte growth phase, protein synthesis on free polysomes is limited by the amount of accessible mRNA. Injecting mRNA into these young oocytes can increase the total protein synthetic rate by approximately three-fold, raising it quantitatively to the level of stage VI oocytes (147). Since the putative mRNA levels are constant throughout these stages, it may be concluded that a large fraction of the maternal mRNA is not available for translation at early stages. This sequestration is gradually relieved so that by stage VI, mRNAs that are translated at all are translated efficiently, and protein synthesis is limited by the translational apparatus; mRNA injected into a stage VI oocyte is translated at the expense of endogenous templates (148). The situation is somewhat different for mRNAs translated on membrane-bound polysomes. This compartment is limited by translational machinery throughout oogenesis, a constraint that can be artificially relaxed by injection of rough endoplasmic reticulum (RER) or salt extracts thereof (149).

The repression of translation in early oocytes is probably due to binding of repressor proteins to mRNA molecules (150,151). Crawford and Richter (152) have characterized a putative example of such repressors. They used a monoclonal antibody to p56, a ribonucleoprotein, to purify RNA associated with this protein. A probe made from this RNA was used to screen a cDNA library, resulting in the identification of several clones corresponding to individual mRNAs that are bound to p56 in early oocytes. It appears that p56 dissociates from these templates concomitantly with the onset of their efficient translation. Further analysis of this interaction may provide insights as to mechanisms of stable translational repression.

In contrast to the general liberation of mRNAs during the final stage of oocyte growth, there are several examples of mRNAs that are maintained in a stored, repressed state until some time during embryonic development. Examples of these include the mRNAs for *c-myc* (153), fibronectin (154), histones (155), lamin (156) and cytoskeletal actins (157). Dworkin and colleagues (158) isolated several cDNA clones corresponding to mRNAs that enter the polysomal fraction at various times during cleavage and Ballantine *et al.* (159) visualized several unidentified polypeptides on two-dimensional gels that fall into this category. It is not known how the translation of these mRNAs is suppressed, but the involvement of proteins analogous to p56 seems likely. Differential translational regulation is a potentially very important level of gene regulation during early *Xenopus* development; particularly as there is little RNA synthesis during the blastula period, a time when crucial inductive interactions are taking place.

A family of proteins whose genes and mRNAs are under multiple levels of control is the collection of approximately 60 polypeptides making up the large and small ribosomal subunits. The mRNAs for ribosomal proteins (RPs) are present in small oocytes and translated with unusually high efficiency; as much as 20% of total protein synthesis at stage III is devoted to making ribosomes (160–162). As most other mRNAs increase in translational efficiency during oogenesis, RP mRNA translation continues at the early oocyte level, and consequently by stage VI these proteins represent a rather small fraction of total synthesis (162). Most RP mRNAS are released from polysomes, deadenylated and degraded after oocyte maturation (163), then begin accumulating during gastrulation, but are not translated until stage 30 (162). There are four exceptions to this pattern. Three RPs, S3, L17 and L31, are synthesized continuously, so their mRNAs must avoid destruction following maturation, while L5 starts to be made at stage 8.5 (164). RPs produced in excess of what can be assembled with rRNA are selectively degraded. In addition, there appears to be a mechanism whereby some unassembled RPs can enter the nucleus and interfere with the removal of introns from their respective mRNA precursor molecules (165).

5. Prospects for studies of the molecular biology of *Xenopus* development

5.1 Disruption of development with sense and antisense reagents

Xenopus is not a convenient subject for extensive genetic screening; the breeding cycle is 1 year or more and it is relatively difficult to sort and maintain more than a few hundred individuals in a laboratory. Fortunately, the capacity of the *Xenopus* oocyte and fertilized egg to incorporate exogenous nucleic acid may provide a partial substitute for mutational analysis. Antisense RNA injected into oocytes can, at least in some cases, base pair with homologous mRNA and block its translation (166). Unfortunately, an abundant endogenous activity appears after fertilization that unwinds RNA–RNA duplexes (167,168), preventing the application of this strategy in the early embryo. This unwinding activity may decline by gastrula or neurula stages, however, opening the possibility of producing antisense RNA in embryos via an injected anti-gene driven by a strong promoter. Giebelhaus *et al.* (169) have used such an approach to inhibit the accumulation of a membrane skeletal protein called p4.1.

This resulted in several abnormalities in affected tadpoles. Specific maternal mRNAs can be ablated by injecting antisense oligodeoxyribonucleotides into oocytes. Following base pairing, an endogenous ribonuclease H activity cleaves the hybridized mRNA, eliminating it from the maternal mRNA pool and thus depriving the cleaving embryo of the encoded protein (170). In instances when a maternal mRNA encodes an essential function and is not translated until after fertilization, such an approach could result in recognizable loss-of-function defects in embryogenesis.

Another way to alter the pattern of gene expression in the early embryo is to inject mRNA into the egg or into selected blastomeres. This has been used to artificially change the levels of the transcription factor TFIIIA, disturbing the regulation of 5S RNA gene expression (126; see Section 4.1). An example of how this approach can be used to disturb morphogenesis is the work of Harvey and Melton (171) with a *Xenopus* homeobox-containing mRNA called Xhox-1A. This mRNA has a homeobox of the *Antennapedia* class, is expressed transiently beginning at midgastrula, and is localized primarily in the axial mesoderm. Synthetic Xhox-1A mRNA injected into one blastomere at the two-cell stage persisted through gastrulation, which is the period when somites become organized into their characteristic periodic pattern. Only one side of the embryo received the injected mRNA, and on that side somite organization was extensively disrupted. Elaborate controls were carried out, including the injection of an Xhox-1A mRNA from which the homeobox had been deleted, to rule out artifactual interference with embryogenesis. This work represents the only published experimental evidence that indicates an important function for homeobox genes in vertebrate embryogenesis, and serves as an example of the potential usefulness of this approach.

5.2 Concluding remarks: transcription factors and developmental biology

Differentiation depends on stable changes in gene expression patterns. These changes are presumably mediated by the activity of protein factors that modulate the transcription, processing or translation of specific genes. Provided the availability of a functional assay, such as gene transfer by microinjection, the DNA sequence elements that interact with regulatory factors can be identified by cloning regulated genes or RNAs, then subjecting these clones to deletion/function analysis, such as was described in Section 4.2. Sites in RNA molecules that interact with regulatory factors should be identifiable by applying similar approaches. When such elements have been sufficiently defined, they can be used to detect the presence, and eventually purify the *trans*-acting factors that bind to them. There has been very encouraging progress in recent years in this area. Synthetic DNA binding sites have been used as the basis for affinity chromatographic purification of class II transcription factors (172 – 176), and have also been used successfully as probes to screen an expression cDNA library (177). In *Xenopus* oocytes it has been possible to activate specific transcriptional regulatory elements by injecting the DNA corresponding to the regulated gene into the oocyte germinal vesicle, and mRNA from expressing cells into the cytoplasm (178). Presumably the injected mRNA includes species that are translated into the appropriate *trans*-acting factors. In principle it should be possible to use this complementation effect as the basis of a strategy for cloning transcription factors. Oocyte injection has also been used to demonstrate the activity of purified transcription factors (for example, see ref. 176).

While there may be special difficulties associated with using embryological material, it seems reasonable to expect that some of these approaches to factor purification will be applicable to genes regulated in early *Xenopus* development. The isolation of such factors, for instance those that control the genes encoding GS17, α-actin and epidermal keratin, ought to provide an experimental foundation for the molecular analysis of how induction and other developmental switches lead to the changes in gene expression that in turn control the formation of the embryo.

6. Acknowledgements

I wish to thank Drs Igor Dawid, Alison Snape and Alan Wolffe for their critical comments and suggestions.

7. References

1. Dumont,J.N. (1972) Oogenesis in *Xenopus laevis* (Daudin). I. Stages of oocyte development in laboratory maintained animals. *J. Morphol.,* **136**, 155–179.
2. Wasserman,W.J., Penna,M.J. and Houle,J.F. (1986) The regulation of *Xenopus laevis* oocyte maturation. In *Gametogenesis and the Early Embryo.* Alan R.Liss, New York, pp. 111–130.
3. Brun,R. (1975) Oocyte maturation in vitro: contribution of the oviduct to total maturation in *Xenopus laevis. Experientia,* **31**, 1275–1276.
4. Elinson,R.P. (1980) The amphibian egg cortex in fertilization and early development. *Symp. Sec. Dev. Biol.,* **38**, 217–234.
5. Gerhart,J., Ubbels,G., Black,S., Hara,K. and Kirschner,M. (1981) A reinvestigation of the role of the grey crescent in axis formation in *Xenopus laevis. Nature,* **292**, 511–516.
6. Palacek,J., Ubbels,G.A. and Rzehak,K. (1978) Changes of the external and internal pigment pattern upon fertilization in the egg of *Xenopus laevis. J. Embryol. Exp. Morphol.,* **45**, 203–214.
7. Newport,J. and Kirshner,M. (1982) A major developmental transition in early *Xenopus* embryos: I. Characterization and timing of cellular changes at the midblastula stage. *Cell,* **30**, 675–686.
8. Kimelman,D., Kirschner,M. and Scherson,T. (1987) The events of the midblastula transition in *Xenopus* are regulated by changes in the cell cycle. *Cell,* **48**, 399–407.
9. Nakakura,N., Miura,T., Yamana,K., Ito,A. and Shiokawa,K. (1987) Synthesis of heterogeneous mRNA-like RNA and low-molecular-weight RNA before the midblastula transition in embryos of *Xenopus laevis. Dev. Biol.,* **123**, 421–429.
10. Signoret,J. and Lefresne,J. (1971) Contribution a l'etude de la segmentation de l'oeuf

d'axolotl. I. Definition de la transition blastuleenne. *Ann. Embryol. Morphogen.*, **4**, 113–123.

11. Gerhart,J.C. (1980) Mechanisms regulating pattern formation in the amphibian egg and early embryo. In *Biological Regulation and Development.* Goldberger,R.F. (ed.), Plenum, New York, Vol. 2, pp. 133–316.

12. Nieuwkoop,P.D. and Faber,J. (1967) *Normal Table of Xenopus laevis* (Daudin). Amsterdam North-Holland.

13. Gerhart,J. and Keller,R. (1986) Region-specific cell activities in amphibian gastrulation. *Annu. Rev. Cell Biol.*, **2**, 201–229.

14. Keller,R. and Danilchik,M. (1988) Regional expression, pattern and timing of convergence and extension during gastrulation of *Xenopus laevis. Development*, **103**, 193–209.

15. Vincent,J.-P., Oster,G.F. and Gerhart,J.C. (1986) Kinematics of grey crescent formation in *Xenopus* eggs: the displacement of subcortical cytoplasm relative to the egg surface. *Dev. Biol.*, **113**, 484–500.

16. Slack,J.M.W. (1983) *From Egg to Embryo. Determinative Events in Early Development.* Cambridge University Press, Cambridge.

17. Steinert,P.M. and Parry,D.A.D. (1985) Intermediate filaments: conformity and diversity of expression and structure. *Annu. Rev. Cell Biol.*, **1**, 41–65.

18. McKeon,F.D., Kirschner,M.W. and Caput,D. (1986) Homologies in both primary and secondary structure between nuclear envelope and intermediate filament proteins. *Nature*, **319**, 463–468.

19. Franke,W.W. (1987) Nuclear lamins and cytoplasmic intermediate filament proteins: a growing multigene family. *Cell*, **48**, 3–4.

20. Godsave,S.F., Wylie,C.C., Lane,E.B. and Anderton,B.H. (1984) Intermediate filaments in the *Xenopus* oocyte: the appearance and distribution of cytokeratin-containing filaments. *J. Embryol. Exp. Morphol.*, **83**, 157–167.

21. Gall,L., Picheral,B. and Gounon,P. (1983) Cytochemical evidence for the presence of intermediate filaments and microfilaments in the egg of *Xenopus laevis. Biol. Cell*, **47**, 331–342.

22. Franz,J.K. and Franke,W.W. (1986) Cloning of cDNA and amino acid sequence of a cytokeratin expressed in oocytes of *Xenopus laevis. Proc. Natl. Acad. Sci. USA*, **83**, 6475–6479.

23. Klymkowsky,M.W., Maynell,L.A. and Polson,A.G. (1987) Polar asymmetry in the organization of the cortical cytokeratin system of *Xenopus laevis* oocytes and embryos. *Development*, **100**, 543–557.

24. Godsave,S.F., Anderton,B.H., Heasman,J. and Wylie,C.C. (1984) Oocytes and early embryos of *Xenopus laevis* contain intermediate filaments which react with anti-mammalian vimentin antibodies. *J. Embryol. Exp. Morphol.*, **83**, 169–187.

25. Tang,P., Sharpe,C.R., Mohun,T.J. and Wylie,C.C. (1988) Vimentin expression in oocytes, eggs and early embryos of *Xenopus laevis. Development*, **103**, 279–287.

26. Jones,E.A. and Woodland,H.R. (1986) Development of the ectoderm in *Xenopus*: tissue specification and the role of cell association and division. *Cell*, **44**, 345–355.

27. Chung,H.-M. and Malacinski,G.M. (1983) Reversal of developmental competence in inverted amphibian eggs. *J. Embryol. Exp. Morphol.*, **73**, 207–220.

28. Moody,S.A. (1987) Fates of the blastomeres of the 16-cell stage *Xenopus* embryo. *Dev. Biol.*, **119**, 560–578.

29. Dale,L. and Slack,J.M.W. (1987) Fate map for the 32-cell stage of *Xenopus laevis. Development*, **99**, 527–551.

30. Miyahara,K., Shiokawa,K. and Yamana,K. (1982) Cellular commitment for post-gastrular increase in alkaline phosphatase activity in *Xenopus laevis* development. *Differentiation*, **21**, 45–49.

31. Smith,L.D. and Williams,M. (1979) Germinal plasm and germ cell determinants in anuran amphibians. In *Maternal Effects in Development.* Newth,D. and Balls,M. (eds), Cambridge University Press, Cambridge, pp.

32. Ikenishi,K. (1987) Functional gametes derived from explants of single blastomeres containing the 'germ plasm' in *Xenopus laevis*: a genetic marker study. *Dev. Biol.*, **122**, 35–38.

33. Smith,L.D. (1966) The role of a 'germinal plasm' in the formation of primordial germ cells in *Rana pipiens. Dev. Biol.*, **14**, 330–347.

34. Holwill,S., Heasman,J., Crawley,C.R. and Wylie,C.C. (1987) Axis and germ line deficiencies caused by u.v. irradiation of *Xenopus* oocytes cultured *in vitro*. *Development*, **100**, 735 – 743.

35. Ikenishi,K., Nakazato,S. and Okuda,T. (1986) Direct evidence for the presence of germ cell determinant in vegetal pole cytoplasm of *Xenopus laevis* and in a subcellular fraction of it. *Dev. Growth Differ.*, **28**, 563 – 568.

36. Wylie,C.C., Holwill,S., O'Driscoll,M., Snape,A. and Heasman,J. (1985) Germ plasm and germ cell determination in *Xenopus laevis* as studied by cell transplantation analysis. *Cold Spring Harbor Symp. Quant. Biol.*, **50**, 37 – 43.

37. Ikenishi,K. and Tsuzaki,Y. (1988) The positional effect of presumptive primordial germ cells (pPGCs) on their differentiation into PGCs in *Xenopus*. *Development*, **102**, 527 – 535.

38. Sutasurya,L.A. and Nieuwkoop,P.D. (1974) The induction of the primordial germ cells in the urodeles. *Roux's Archiv.*, **175**, 199 – 220.

39. Carpenter,C.D. and Klein,W.H. (1982) A gradient of poly(A)$^+$ RNA sequences in *Xenopus laevis* eggs and embryos. *Dev. Biol.*, **91**, 43 – 49.

40. King,M.L. and Barklis,E. (1985) Regional distribution of maternal messenger RNA in the amphibian oocyte. *Dev. Biol.*, **112**, 203 – 212.

41. Rebagliati,M.R., Weeks,D.L., Harvey,R.P. and Melton,D.A. (1985) Identification and cloning of localized maternal RNAs from *Xenopus* eggs. *Cell*, **42**, 769 – 777.

42. Weeks,D.L. and Melton,D.A. (1987) A maternal mRNA localized to the animal pole of *Xenopus* eggs encodes a subunit of mitochondrial ATPase. *Proc. Natl. Acad. Sci. USA*, **84**, 2798 – 2802.

43. Weeks,D.L. and Melton,D.A. (1987) A maternal mRNA localized to the vegetal hemisphere in *Xenopus* eggs codes for a growth factor related to TGF-β. *Cell*, **51**, 861 – 867.

44. Padgett,R.W., St Johnston,R.D. and Gelbart,W.M. (1987) A transcript from a *Drosophila* pattern gene predicts a protein homologous to the transforming growth factor-β family. *Nature*, **325**, 81 – 84.

45. Cheifetz,S., Weatherbee,J.A., Tsang,M.L.-S., Anderson,J.K., Mole,J.E., Lucas,R. and Massague,J. (1987) The transforming growth factor-β system, a complex pattern of cross-reactive ligands and receptors. *Cell*, **48**, 409 – 415.

46. Sporn,M.B., Roberts,A.B., Wakefield,L.M. and de Crombrugghe,B. (1987) Some recent advances in the chemistry and biology of transforming growth factor-beta. *J. Cell Biol.*, **105**, 1039 – 1045.

47. Gerhart,J.C., Vincent,J.-P., Scharf,S.R., Black,S.D., Gimlich,R.L. and Danilchik,M. (1984) Localization and induction in early development of *Xenopus*. *Phil. Trans. R. Soc. Lond. B*, **307**, 319 – 330.

48. Vincent,J.-P. and Gerhart,J.C. (1987) Subcortical rotation in *Xenopus* eggs: an early step in embryonic axis specification. *Dev. Biol.*, **123**, 526 – 539.

49. Elinson,R.P. (1983) Cytoplasmic phases in the first cell cycle of the activated frog egg. *Dev. Biol.*, **100**, 440 – 451.

50. Elinson,R.P. (1985) Changes in levels of polymeric tubulin associated with activation and dorsoventral polarization of the frog egg. *Dev. Biol.*, **109**, 224 – 233.

51. Gerhart,J., Danilchik,M., Roberts,J., Rowning,B. and Vincent,J.-P. (1986) Primary and secondary polarity of the amphibian oocyte and egg. In *Gametogenesis and the Early Embryo*. pp. 305 – 319.

52. Grant,P. and Wacaster,J.F. (1972) The amphibian grey crescent—a site of developmental information? *Dev. Biol.*, **28**, 454 – 471.

53. Malacinski,G.M., Benford,H. and Chung,H.-M. (1975) Association of an ultraviolet irradiation sensitive cytoplasmic localization with the future dorsal side of the amphibian egg. *J. Exp. Zool.*, **191**, 97 – 110.

54. Malacinski,G.M., Brothers,A.J. and Chung,H.-M. (1977) Destruction of components of the neural induction system of the amphibian egg with ultraviolet irradiation. *Dev. Biol.*, **56**, 24 – 39.

55. Scharf,S.R. and Gerhart,J.C. (1980) Determination of the dorsal – ventral axis in eggs of *Xenopus laevis*: complete rescue of UV-impaired eggs by oblique orientation before first cleavage. *Dev. Biol.*, **79**, 181 – 198.

56. Cleine,J.H. (1986) Replacement of posterior by anterior endoderm reduces sterility

in embryos from inverted eggs of *Xenopus laevis*. *J. Embryol. Exp. Morphol.*, **94**, 83–93.

57. Scharf,S.R., Vincent,J.-P. and Gerhart,J.C. (1984) Axis determination in the *Xenopus* egg. In *Molecular Biology of Development*. UCLA Symposia on Molecular and Cellular Biology, Volume 19, Alan R.Liss, New York, pp.51–73.

58. Scharf,S.R. and Gerhart,J.C. (1983) Axis determination in eggs of *Xenopus laevis*: a critical period before first cleavage, identified by the common effects of cold, pressure and ultraviolet irradiation. *Dev. Biol.*, **99**, 75–87.

59. Gerhart,J., Black,S., Roberts,J., Rowning,B., Scharf,S. and Vincent,J.-P. (1987) Localized activation of dorsal development in *Xenopus* eggs. In *Molecular Approaches to Developmental Biology*, UCLA Symposia on Molecular and Cellular Biology, Alan R.Liss, New York, pp. 89–95.

60. Smith,J.C. (1987) A mesoderm-inducing factor is produced by a *Xenopus* cell line. *Development*, **99**, 3–14.

61. Sargent,T.D., Jamrich,M. and Dawid,I.B. (1986) Cell interactions and the control of gene activity during early development of *Xenopus laevis*. *Dev. Biol.*, **114**, 238–246.

62. Sudarwati,S. and Nieuwkoop,P.D. (1971) Mesoderm formation in the anuran *Xenopus laevis* (Daudin). *Roux's Archiv.*, **166**, 189–204.

63. Boterenbrood,E.C. and Nieuwkoop,P.D. (1973) The formation of the mesoderm in Urodelean amphibians. V. Its regional induction by the endoderm. *Roux's Archiv.*, **173**, 319–332.

64. Dale,L. and Slack,J.M.W. (1987) Regional specification with the mesoderm of early embryos of *Xenopus laevis*. *Development*, **100**, 279–295.

65. Smith,J.C., Yaqoob,M. and Symes,K. (1988) Purification, partial characterization and biological effects of the XTC mesoderm-inducing factor. *Development*, **103**, 591–600.

66. Slack,J.M.W., Darlington,B.G., Heath,J.K. and Godsave,S.F. (1987) Mesoderm induction in early *Xenopus* embryos by heparin-binding growth factors. *Nature*, **326**, 197–200.

67. Slack,J.M.W., Isaacs,H.V. and Darlington,B.G. (1988) Inductive effects of fibroblast growth factor and lithium ion on *Xenopus* blastula ectoderm. *Development*, **103**, 581–590.

68. Jones,E.A. and Woodland,H.R. (1987) The development of animal cap cells in *Xenopus*: a measure of the start of animal cap competence to form mesoderm. *Development*, **101**, 557–563.

69. Gurdon,J.B., Fairman,S., Mohun,T.J. and Brennan,S. (1985) Activation of muscle-specific actin genes in *Xenopus* development by an induction between animal and vegetal cells of a blastula. *Cell*, **41**, 913–922.

70. Gimlich,R.L. (1986) Acquisition of developmental autonomy in the equatorial region of the *Xenopus* embryo. *Dev. Biol.*, **115**, 340–352.

71. Nakamura,O., Takasaki,H. and Mizohata,T. (1970) Differentiation during cleavage in *Xenopus laevis*. I. Acquisition of self-differentiation capacity of the dorsal marginal zone. *Proc. Jap. Acad.*, **46**, 694–699.

72. Gurdon,J.B., Mohun,T.J., Fairman,S. and Brennan,S. (1985) All components required for the eventual activation of muscle-specific actin genes are localized in the subequatorial region of an uncleaved amphibian egg. *Proc. Natl. Acad. Sci. USA*, **82**, 139–143.

73. Kao,K.R., Masui,Y. and Elinson,R.P. (1986) Lithium-induced respecification of pattern in *Xenopus laevis* embryos. *Nature*, **322**, 371–373.

74. Render,J.A. and Elinson,R.P. (1986) Axis determination in polyspermic *Xenopus laevis* eggs. *Dev. Biol.*, **115**, 425–433.

75. Kageura,H. and Yamana,K. (1983) Pattern regulation in isolated halves and blastomeres of early *Xenopus laevis*. *J. Embryol. Exp. Morphol.*, **74**, 221–234.

76. Kageura,H. and Yamana,K. (1984) Pattern regulation in defect embryos of *Xenopus laevis*. *Dev. Biol.*, **101**, 410–415.

77. Cooke,J. (1985) Early specification for body position in mes-endodermal regions of an amphibian embryo. *Cell Differ.*, **17**, 1–12.

78. Cardellini,P. (1988) Reversal of dorsoventral polarity in *Xenopus laevis* embryos by 180° rotation of the animal micromeres at the eight-cell stage. *Dev. Biol.*, **128**, 428–434.

79. Gimlich,R.L. and Gerhart,J.C. (1984) Early cellular interactions promote embryonic

axis formation in *Xenopus laevis. Dev. Biol.,* **104**, 117 – 130.

80. Slack,J.M.W. and Forman,D. (1980) An interaction between dorsal and ventral regions of the marginal zone in early amphibian embryos. *J. Embryol. Exp. Morphol.,* **56**, 283 – 299.
81. Curtis,A.S.G. (1960) Cortical grafting in *Xenopus laevis. J. Embryol. Exp. Morphol.,* **8**, 163 – 173.
82. Curtis,A.S.G. (1962) Morphogenetic interactions before gastrulation in the amphibian, *Xenopus laevis*—the cortical field. *J. Embryol. Exp. Morphol.,* **10**, 410 – 422.
83. Tomkins,R. and Rodman,W.P. (1971) The cortex of *Xenopus laevis* embryos: regional differences in composition and biological activity. *Proc. Natl. Acad. Sci. USA,* **68**, 2921 – 2923.
84. Kintner,C.R. and Melton,D.A. (1987) Expression of *Xenopus* N-CAM RNA in ectoderm is an early response to neural induction. *Development,* **99**, 311 – 325.
85. Sharpe,C.R. (1988) Developmental expression of a neurofilament-M and two vimentin-like genes in *Xenopus laevis. Development,* **103**, 269 – 277.
86. Sharpe,C.R., Fritz,A., De Robertis,E.M. and Gurdon,J.B. (1987) A homeobox-containing marker of posterior neural differentiation shows the importance of predetermination in neural induction. *Cell,* **50**, 749 – 758.
87. Jacobson,M. and Rutishauser,U. (1986) Induction of neural cell adhesion molecule (NCAM) in *Xenopus* embryos. *Dev. Biol.,* **116**, 524 – 531.
88. Jamrich,M., Sargent,T.D. and Dawid,I.B. (1987) Cell-type-specific expression of epidermal cytokeratin genes during gastrulation of *Xenopus laevis. Genes Dev.,* **1**, 124 – 132.
89. Jacobson,A.G. (1966) Inductive processes in embryonic development. *Science,* **152**, 25 – 34.
90. Henry,J.J. and Grainger,R.M. (1987) Inductive interactions in the spatial and temporal restriction of lens-forming potential in embryonic ectoderm of *Xenopus laevis. Dev. Biol.,* **124**, 200 – 214.
91. Grainger,R.M., Henry,J.J. and Henderson,R.A. (1988) Reinvestigation of the role of the optic vesicle in embryonic lens induction. *Development,* **102**, 517 – 526.
92. Grunz,H., McKeehan,W.L., Knochel,W., Born,J., Tiedemann,H. and Tiedemann,H. (1988) Induction of mesodermal tissues by acidic and basic heparin binding growth factors. *Cell Differ.,* **22**, 183 – 190.
93. Kimelman,D. and Kirschner,M. (1987) Synergistic induction of mesoderm by FGF and TGFβ and the identification of an mRNA coding for FGF in the early *Xenopus* embryo. *Cell,* **51**, 869 – 877.
94. Rosa,F., Roberts,A.B., Danielpour,D., Dart,L.L., Sporn,M.B. and Dawid,I.B. (1988) Mesoderm induction in amphibians: the role of TGF-β2-like factors. *Science,* **239**, 783 – 785.
95. Kao,K.R. and Elinson,R.P. (1988) The entire mesodermal mantle behaves as Spemann's organizer in dorsoanterior enhanced *Xenopus laevis* embryos. *Dev. Biol.,* **127**, 64 – 77.
96. Altman,J. (1988) Ins and outs of cell signalling. *Nature,* **331**, 119 – 120.
97. Avissar,S., Schreiber,G., Danon,A. and Belmaker,R.H. (1988) Lithium inhibits adrenergic and cholinergic increases in GTP binding in rat cortex. *Nature,* **331**, 440 – 442.
98. Stryer,L. and Bourne,H.R. (1986) G proteins: a family of signal transducers. *Annu. Rev. Cell Biol.,* **2**, 391 – 419.
99. Ignotz,R.A. and Massague,J. (1985) Type β transforming growth factor controls the adipogenic differentiation of 3T3 fibroblasts. *Proc. Natl. Acad. Sci. USA,* **82**, 8530 – 8534.
100. Massague,J., Cheifetz,S., Endo,T. and Nadal-Ginard,B. (1986) Type β transforming growth factor is an inhibitor of myogenic differentiation. *Proc. Natl. Acad. Sci. USA,* **83**, 8206 – 8210.
101. Lathrop,B., Olson,E. and Glaser,L. (1985) Control by fibroblast growth factor of differentiation in the BC3H1 muscle cell line. *J. Cell Biol.,* **100**, 1540 – 1547.
102. Olson,E.N., Spizz,G. and Tainsky,M.A. (1987) The oncogenic forms of N-ras or H-ras prevent skeletal myoblast differentiation. *Mol. Cell. Biol.,* **7**, 2104 – 2111.
103. Yu,C.-L., Tsai,M.-H. and Stacey,D.W. (1988) Cellular ras activity and phospholipid metabolism. *Cell,* **52**, 63 – 71.

104. Fedoroff,N.V. and Brown,D.D. (1978) The nucleotide sequence of oocyte 5S DNA in *Xenopus laevis*. I. The AT-rich spacer. *Cell,* **13**, 701–716.
105. Peterson,R.C., Doering,J.L. and Brown,D.D. (1980) Characterization of two *Xenopus* somatic 5S DNAs and one minor oocyte-specific 5S DNA. *Cell,* **20**, 131–141.
106. McConkey,G.A. and Bogenhagen,D.F. (1987) Transition mutations within the *Xenopus borealis* somatic 5S RNA gene can have independent effects on transcription and TFIIIA binding. *Mol. Cell. Biol.,* **7**, 486–494.
107. Bogenhagen,D.F., Sakonju,S. and Brown,D.D. (1980) A control region in the center of the 5S RNA gene directs specific initiation of transcription. II. The 3′ border of the region. *Cell,* **19**, 27–35.
108. Sakonju,S., Bogenhagen,D.F. and Brown,D.D. (1980) A control region in the center of the 5S RNA gene directs specific initiation of transcription. I. The 5′ border of the region. *Cell,* **19**, 13–25.
109. Lassar,A.B., Martin,P.L. and Roeder,R.G. (1983) Transcription of class III genes: formation of preinitiation complexes. *Science,* **222**, 740–748.
110. Pieler,T., Hamm,J. and Roeder,R.G. (1987) The 5S gene internal control region is composed of three distinct sequence elements, organized as two functional domains with variable spacing. *Cell,* **48**, 91–100.
111. Peck,L.J., Millstein,K., Eversole-Cire,P., Gottesfeld,J.M. and Varshavsky,A. (1987) Transcriptionally inactive oocyte-type 5S RNA genes of *Xenopus laevis* are complexed with TFIIIA *in vitro*. *Mol. Cell. Biol.,* **7**, 3503–3510.
112. Reynolds,W.F. and Azer,K. (1988) Sequence differences upstream of the promoters are involved in the differential expression of the *Xenopus* somatic and oocyte 5S RNA genes. *Nucleic Acids Res.,* **16**, 3391–3403.
113. Picard,B. and Wegnez,M. (1979) Isolation of a 7S particle from *Xenopus laevis* oocytes: a 5S RNA–protein complex. *Proc. Natl. Acad. Sci. USA,* **76**, 241–245.
114. Honda,B.M. and Roeder,R.G. (1980) Association of a 5S gene transcription factor with 5S RNA and altered levels of the factor during cell differentiation. *Cell,* **22**, 119–126.
115. Pelham,H.R.B. and Brown,D.D. (1980) A specific transcription factor that can bind either the 5S RNA gene or 5S RNA. *Proc. Natl. Acad. Sci. USA,* **77**, 4170–4174.
116. Ginsberg,A.M., King,B.O. and Roeder,R.G. (1984) *Xenopus* 5S gene transcription factor TFIIIA: characterization of a cDNA clone and measurement of RNA levels throughout development. *Cell,* **39**, 479–489.
117. Miller,J., McLachlan,A.D. and Klug,A. (1985) Repetitive zinc-binding domains in the protein transcription factor IIIA from *Xenopus* oocytes. *EMBO J.,* **4**, 1609–1614.
118. Wolffe,A.P. (1988) Transcription fraction TFIIIC can regulate differential *Xenopus* 5S RNA gene transcription *in vitro*. *EMBO J.,* **7**, 1071–1079.
119. Setzer,D.R. and Brown,D.D. (1985) Formation and stability of the 5S RNA transcription complex. *J. Biol. Chem.,* **260**, 2483–2492.
120. Bieker,J.J., Martin,P.L. and Roeder,R.G. (1985) Formation of a rate-limiting intermediate in 5S RNA gene transcription. *Cell,* **40**, 119–127.
121. Bogenhagen,D.F., Wormington,W.M. and Brown,D.D. (1982) Stable transcription complexes of *Xenopus* 5S RNA genes: a means to maintain the differentiated state. *Cell,* **28**, 413–421.
122. Wolffe,A.P., Jordan,E. and Brown,D.D. (1986) A bacteriophage RNA polymerase transcribes through a *Xenopus* 5S RNA gene transcription complex without disrupting it. *Cell,* **44**, 381–389.
123. Wolffe,A.P. and Brown,D.D. (1986) DNA replication *in vitro* erases a *Xenopus* 5S RNA gene transcription complex. *Cell,* **47**, 217–227.
124. Brown,D.D. (1984) The role of stable complexes that repress and activate eucaryotic genes. *Cell,* **37**, 359–365.
125. Wolffe,A.P. and Brown,D.D. (1987) Differential 5S RNA gene expression *in vitro*. *Cell,* **51**, 733–740.
126. Andrews,M.T. and Brown,D.D. (1987) Transient activation of oocyte 5S RNA genes in *Xenopus* embryos by raising the level of the *trans*-acting factor TFIIIA. *Cell,* **51**, 445–453.
127. McConkey,G.A. and Bogenhagen,D.F. (1988) TFIIIA binds with equal affinity to somatic and major oocyte 5S RNA genes. *Genes Dev.,* **2**, 205–214.

128. Schlissel,M.S. and Brown,D.D. (1984) The transcriptional regulation of *Xenopus* 5S RNA genes in chromatin: the roles of active stable transcription complexes and histone H1. *Cell,* **37**, 903–913.

129. Korn,L.J., Guinta,D.R. and Tso,J.Y. (1987) TFIIIA mediated control of 5S RNA gene expression in *Xenopus*. In *Molecular Approaches to Developmental Biology,* UCLA Symposia on Molecular and Cellular Biology, Alan R.Liss, New York, pp. 25–37.

130. Etkin,L.D. and Pearman,B. (1987) Distribution, expression and germ line transmission of exogenous DNA sequences following microinjection into *Xenopus laevis* eggs. *Development,* **99**, 15–23.

131. Laskey,R.A., Mills,A.D. and Morris,N.R (1977) Assembly or SV40 chromatin in a cell-free system from *Xenopus* eggs. *Cell,* **10**, 237–243.

132. Forbes,D.J., Kirschner,M.W. and Newport,J.W. (1983) Spontaneous formation of nucleus-like structures around bacteriophage DNA microinjected in *Xenopus* eggs. *Cell,* **34**, 13–23.

133. Lohka,M.J. and Masui,Y. (1983) Formation *in vitro* of sperm pronuclei and mitotic chromosomes induced by amphibian ooplasmic components. *Science,* **220**, 719–721.

134. Blow,J.J. and Laskey,R.A. (1986) Initiation of DNA replication in nuclei and purified DNA by a cell-free extract of *Xenopus* eggs. *Cell,* **47**, 577–587.

135. Newport,J. (1987) Nuclear reconstitution *in vitro*: stages of assembly around protein-free DNA. *Cell,* **48**, 205–217.

136. Bendig,M.M. and Williams,J.G. (1984) Differential expression of the *Xenopus laevis* tadpole and adult β-globin genes when injected into fertilized *Xenopus laevis* eggs. *Mol. Cell. Biol.,* **4**, 567–570.

137. Krieg,P.A. and Melton,D.A. (1985) Developmental regulation of a gastrula-specific gene injected into fertilized *Xenopus* eggs. *EMBO J.,* **4**, 3463–3471.

138. Krieg,P.A. and Melton,D.A. (1987) An enhancer responsible for activating transcription at the midblastula transition in *Xenopus* development. *Proc. Natl. Acad. Sci. USA,* **84**, 2331–2335.

139. Wilson,C., Cross,G.S. and Woodland,H.R. (1986) Tissue-specific expression of actin genes injected into *Xenopus* embryos. *Cell,* **47**, 589–599.

140. Mohun,T.J., Garrett,N. and Gurdon,J.B. (1986) Upstream sequences required for tissue-specific activation of the cardiac actin gene in *Xenopus laevis* embryos. *EMBO J.,* **5**, 3185–3193.

141. Treisman,R. (1986) Identification of a protein-binding site that mediates transcriptional response of the c-fos gene to serum factors. *Cell,* **46**, 567–574.

142. Greenberg,M.E. and Ziff,E.B. (1984) Stimulation of 3T3 cells induces transcription of the c-fos proto-oncogene. *Nature,* **311**, 433–438.

143. Dawid,I.B., Rebbert,M.L., Rosa,F., Jamrich,M. and Sargent,T.D. (1989) Gene expression in amphibian embryogenesis. In *Regulatory Mechanisms in Developmental Processes*. Eguchi,G., Okuda,T.S. and Saxen,L. (eds), Elsevier, Amsterdam, pp. 67–74.

144. Miyatani,S., Winkles,J.A., Sargent,T.D. and Dawid,I.B. (1986) Stage-specific keratins in *Xenopus laevis* embryos and tadpoles: the XK81 gene family. *J. Cell Biol.,* **103**, 1957–1965.

145. Dolecki,G.J. and Smith,L.D. (1979) Poly(A)$^+$ RNA metabolism during oogenesis in *Xenopus laevis*. *Dev. Biol.,* **69**, 217–236.

146. Golden,L., Schafer,U. and Rosbash,M. (1980) Accumulation of individual pA+ RNAs during oogenesis of *Xenopus laevis*. *Cell,* **22**, 835–844.

147. Taylor,M.A., Johnson,A.D. and Smith,L.D. (1985) Growing *Xenopus* oocytes have spare translational capacity. *Proc. Natl. Acad. Sci. USA,* **82**, 6586–6589.

148. Laskey,R.A., Mills,A.D., Gurdon,J.B. and Partington,G.A. (1977) Protein synthesis in oocytes of *Xenopus laevis* is not regulated by the supply of messenger RNA. *Cell,* **11**, 345–351.

149. Richter,J.D., Evers,D.C. and Smith,L.D. (1983) The recruitment of membrane-bound mRNAs for translation in microinjected *Xenopus* oocytes. *J. Biol. Chem.,* **258**, 2614–2620.

150. Richter,J.D. and Smith,L.D. (1983) Developmentally regulated RNA binding proteins during oogenesis in *Xenopus laevis*. *J. Biol. Chem.,* **258**, 4864–4869.

151. Richter,J.D. and Smith,L.D. (1984) Reversible inhibition of translation by *Xenopus* oocyte-specific proteins. *Nature,* **309**, 378–380.
152. Crawford,D.R. and Richter,J.D. (1987) An RNA-binding factor protein from *Xenopus* oocytes is associated with specific message sequences. *Development,* **101**, 741–749.
153. King,M.W., Roberts,J.M. and Eisenman,R.N. (1986) Expression of the c-myc proto-oncogene during development of *Xenopus laevis. Mol. Cell. Biol.,* **6**, 4499–4508.
154. Lee,G., Hynes,R. and Kirschner,M. (1984) Temporal and spatial regulation of fibronectin in early *Xenopus* development. *Cell,* **36**, 729–740.
155. Flynn,J.M. and Woodland,H.R. (1980) The synthesis of histone H1 during amphibian development. *Dev. Biol.,* **75**, 222–230.
156. Stick,R. and Hausen,P. (1985) Changes in the nuclear lamina composition during early development in *Xenopus laevis. Cell,* **41**, 191–200.
157. Sturgess,E.A., Ballantine,J.E.M., Woodland,H.R., Mohun,P.R., Lane,C.D. and Dimitriadis,G.J. (1980) Actin synthesis during the early development of *Xenopus laevis. J. Embryol. Exp. Morphol.,* **58**, 303–320.
158. Dworkin,M.B., Shrutkowski,A. and Dworkin-Rastl,E. (1985) Mobilization of specific maternal RNA species into polysomes after fertilization in *Xenopus laevis. Proc. Natl. Acad. Sci. USA,* **82**, 7636–7640.
159. Ballantine,J.E.M., Woodland,H.R. and Sturgess,E.A. (1979) Changes in protein synthesis during the development of *Xenopus laevis. J. Embryol. Exp. Morphol.,* **51**, 137–153.
160. Pierandrei-Amaldi,P. and Beccari,E. (1980) Messenger RNA for ribosomal proteins in *Xenopus laevis* oocytes. *Eur. J. Biochem.,* **106**, 603–611.
161. Dixon,L.K. and Ford,P.J. (1982) Regulation of protein synthesis and accumulation during oogenesis in *Xenopus laevis. Dev. Biol.,* **93**, 478–497.
162. Baum,E.Z. and Wormington,W.M. (1985) Coordinate expression of ribosomal protein genes during *Xenopus* development. *Dev. Biol.,* **111**, 488–498.
163. Hyman,L.E. and Wormington,W.M. (1988) Translational inactivation of ribosomal protein mRNAs during *Xenopus* oocyte maturation. *Genes Dev.,* **2**, 598–605.
164. Pierandrei-Amaldi,P., Campioni,N., Beccari,E., Bozzoni,I. and Amaldi,F. (1982) Expression of ribosomal-protein genes in *Xenopus laevis* development. *Cell,* **30**, 163–171.
165. Pierandrei-Amaldi,P., Beccari,E., Bozzoni,I. and Amaldi,F. (1985) Ribosomal protein production in normal and anucleolate *Xenopus* embryos: regulation at the posttranscription and translational levels. *Cell,* **42**, 317–323.
166. Melton,D.A. (1985) Injected anti-sense RNAs specifically block messenger RNA translation *in vivo. Proc. Natl. Acad. Sci. USA,* **82**, 144–148.
167. Bass,B.L. and Weintraub,H. (1987) A developmentally regulated activity that unwinds RNA duplexes. *Cell,* **48**, 607–613.
168. Rebagliati,M.R. and Melton,D.A. (1987) Antisense RNA injections in fertilized frog eggs reveal an RNA duplex unwinding activity. *Cell,* **48**, 599–605.
169. Giebelhaus,D.H., Eib,D.W. and Moon,R.T. (1988) Antisense RNA inhibits expression of membrane skeleton protein 4.1 during embryonic development of *Xenopus. Cell,* **53**, 601–615.
170. Shuttleworth,J. and Colman,A. (1988) Antisense oligonucleotide-directed cleavage of mRNA in *Xenopus* oocytes and eggs. *EMBO J.,* **7**, 427–434.
171. Harvey,R.P. and Melton,D.A. (1988) Microinjection of synthetic Xhox-1A homeobox mRNA disrupts somite formation in developing *Xenopus* embryos. *Cell,* **53**, 687–697.
172. Kadonaga,J.T. and Tjian,R. (1986) Affinity purification of sequence-specific DNA binding proteins. *Proc. Natl. Acad. Sci. USA,* **83**, 5889–5893.
173. Scheidereit,C., Heguy,A. and Roeder,R.G. (1987) Identification and purification of a human lymphoid-specific octamer-binding protein (OTF-2) that activates transcription of an immunoglobulin promoter *in vitro. Cell,* **51**, 783–793.
174. Singh,R.P. and Natarajan,V. (1987) A rapid affinity method for isolation and characterization of sequence specific DNA binding factor. *Biochem. Biophys. Res. Commun.,* **147**, 65–70.
175. Wang,L.H., Tsai,S.Y., Sagami,I., Tsai,M.J. and O'Malley,B.W. (1987) Purification and characterization of chicken ovalbumin upstream promoter transcription factor from HeLa cells. *J. Biol. Chem.,* **262**, 16080–16086.

176. Wu,C., Wilson,S., Walker,B., Dawid,I., Paisley,T., Zimarino,V. and Ueda,H. (1987) Purification and properties of *Drosophila* heat shock activator protein. *Science,* **238**, 1247–1253.

177. Singh,H., LeBowitz,J.H., Baldwin,A.S., Jr and Sharp,P.A. (1988) Molecular cloning of an enhancer binding protein: isolation by screening of an expression library with a recognition site DNA. *Cell,* **52**, 415–423.

178. Sweeney,G.E. and Old,R.W. (1988) *Trans*-activation of transcription, from promoters containing immunoglobulin gene octamer sequences, by myeloma cell mRNA in *Xenopus* oocytes. *Nucleic Acid Res.,* **16**, 4903–4913.

5
The mouse
Ian J.Jackson

1. Introduction

In our anthropocentric way we regard developmental systems as models for human development. By its likeness the mouse is, of course, the most apt of the commonly studied models. It is also the least tractable. Development within the uterus creates a number of problems for the developmental biologist. Experimental manipulation of the embryo after implantation is very difficult, and litter size is relatively small so material at any stage (but particularly the early stages) is hard to come by. The small litter size, fairly long generation time and space needed to maintain the animals means that the genetics of the mouse has been developed mostly through intensive efforts at a few centers and is rather limited when compared to *Drosophila* or *Caenorhabditis*.

The advent of molecular biology on this particular frontier has provided the solution to many technical problems. Moreover, the impetus given to the field as a whole by the technology, and the optimistic views held by its practitioners that all problems are soluble, has resulted in advances in areas in which the impact of molecular biology would not have been thought obvious.

In this chapter I cannot hope to cover all aspects of early development and consequently, it contains largely a reflection of personal biases and selection. Two approaches to the problem can be taken: a molecular approach and a genetic approach. The former uses biochemical or molecular probes as tools to understand the phenomenology of development. The latter uses genetics, which coupled with molecular techniques can give an understanding of how genes function to regulate development.

2. Outline of early mouse development

The process of development begins before fertilization, during oogenesis,
However, maternally encoded mRNA is less important in mouse
development than in *Drosophila*, for example, and virtually all early
development is determined by zygotically transcribed genes. I will
describe the differentiative events of early development, from fertilization
to the point where primary inductive processes have occurred, when the
mesoderm, neurectoderm and somites have formed. A good basic
description of events, and many references can be found in Hogan *et al.* (1).
   The major landmarks after fertilization include:
(i)   the onset of zygotic transcription;
(ii)   compaction, when tight intercellular contacts are made;
(iii)   implantation of the blastocyst into the uterus;
(iv)   gastrulation, when the mesoderm is formed.
The embryo can be cultured *in vitro* from fertilization to the point at which
it normally implants 4.5 days later (2). Beyond this point embryos can
be removed, manipulated and cultured for short periods *in vitro* but have
not been successfully reimplanted.
   Much of the early development of the mouse involves formation of
structures which support the embryo in its intra-uterine environment, the
placenta, yolk sac and amnion. The embryo proper does not begin to
develop until about 7 days post-fertilization (days *post coitum*, dpc),
one-third of the way through gestation. *Figure 1* shows schematically the
major differentiative features of early development. The naming of parts
does not merely describe the position of cells, but implies differentiation
of morphology, of function and of fate.

2.1 Pre-implantation

The oocyte released by the ovary has undergone only part of meiosis.
The first meiotic division has occurred and one diploid set of chromosomes
has been segregated outside the cell in the first polar body. The second
meiotic cleavage has been initiated but does not complete until after
fertilization. The sperm enters the egg opposite the polar body, forms
the male pronucleus and triggers completion of meiosis. The maternal
pronucleus forms, the other haploid chromosome set is eliminated as the
second polar body and the two pronuclei fuse to become the zygotic
nucleus. The cleavage stages which follow are slow; the initial division
takes 24 h, and the next few are at about 12 h intervals. The cleavage
embryo is not yet developmentally restricted. In sheep, blastomeres of
an eight-cell embryo can give rise individually to whole embryos (3). In
mouse, fusion of cells from two eight-cell embryos results in chimeras
in which each donor can contribute to all tissues, including extraembryonic
ones (4,5).

At the eight-cell stage the cells compact to form a tighter aggregation (the morula), when gap junctions form between cells. Compaction establishes polarity in the cells, producing an inside and an outside face (6). This intracellular polarity gives rise, on subsequent cell division, to intercellular differences (inside and outside cells), which may underlie the differentiation of the two cell types of the blastocyst. Tight cell connections around the outside allow fluid to accumulate by an active process in the blastocoel cavity. Cells on the inside become the inner cell mass (ICM) and those surrounding it and the blastocoel become the trophectoderm. At this stage the embryo has about 64 cells, and has not increased in mass or volume since fertilization. Indeed, it is still enclosed in the zona pellucida, the coat which surrounds the oocyte. The two cell types in the blastocyst can be physically separated, and shown to be largely developmentally restricted to their own separate lineages.

Before implantation two other differentiative events occur. Trophectoderm cells differentiate, according to position, into mural trophectoderm in contact with the blastocoel, and polar trophectoderm above the ICM. The mural cells do not divide further, but continue DNA synthesis and become polyploid 'giant cells'. Trophoblasts from which the ICM is removed become giant cells, indicating the instructive role of the ICM in determining polar trophoblast. The ICM meantime also undergoes differentiation. Cells in contact with the blastocoel organize into a layer known as primitive endoderm, which surrounds a core of primitive ectoderm, the epiblast. The embryo proper develops solely from the epiblast, whilst the trophoblast and primitive endoderm derivatives (with some primitive ectoderm derivatives) go on to form the yolk sacs, amnion and placenta.

Those concerned with the roots of words will note that the outer cell layer here has the root applied to an internal feature (endo) and the core has the external (ecto) root. This is because the development of mouse and rat embryos is different from almost all other mammals; in others the embryo develops not as a cup, but as a disc, in which the ectoderm sits on top of the endoderm.

2.2 Post-implantation

Just before implantation at about 4.5 dpc the blastocyst hatches from the zona pellucida which surrounds it. The mural trophoblast contacts the uterine wall and penetrates the epithelial layer. A proliferation of uterine cells is induced at the site of implantation.

The egg cylinder forms as a result of polar trophectoderm differentiating to become the ectoplacental cone and, below it and pushing the epiblast down into the blastocoel, the extraembryonic ectoderm. The primitive endoderm around the epiblast proliferates to surround the egg cylinder and to line the inner face of the mural trophectoderm.

This endoderm differentiates into visceral endoderm (around the egg cylinder) and parietal endoderm (on the blastocoel face of the mural trophectoderm) where they contribute to the visceral and parietal yolk sacs. The parietal endoderm cells synthesize and lay down a basement membrane (Reichert's membrane) which separates them from the giant cells and encloses the embryo. Visceral endoderm synthesizes many proteins later made by the embryonic liver and is later joined at gastrulation by mesodermal cells which form hematopoeitic blood islands in the visceral yolk sac.

Around this time a cavity, the proamniotic cavity, forms inside the epiblast and expands to fill both the embryonic and extraembryonic part of the egg cylinder, which thus becomes a bilayered cup of ectoderm enclosed by endoderm.

2.3 Gastrulation

About 7 days post-fertilization, the mesoderm forms between the two layers of the egg cylinder. Cells separate from the ectoderm, beginning at what will become the posterior end. Some move away from the embryo in a fold across the proamniotic cavity which encloses a new cavity, the

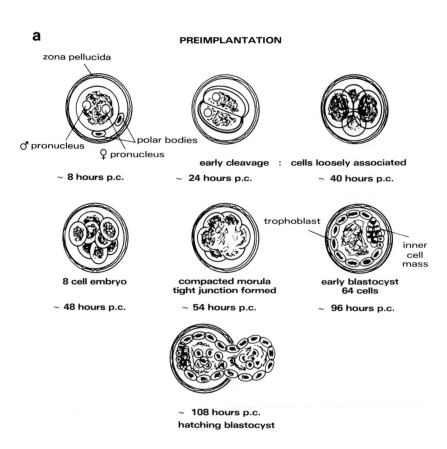

a   PREIMPLANTATION

zona pellucida

♂ pronucleus
♀ pronucleus   polar bodies

~ 8 hours p.c.

~ 24 hours p.c.
early cleavage : cells loosely associated

~ 40 hours p.c.

trophoblast

inner cell mass

8 cell embryo

~ 48 hours p.c.

compacted morula
tight junction formed

~ 54 hours p.c.

early blastocyst
64 cells

~ 96 hours p.c.

~ 108 hours p.c.
hatching blastocyst

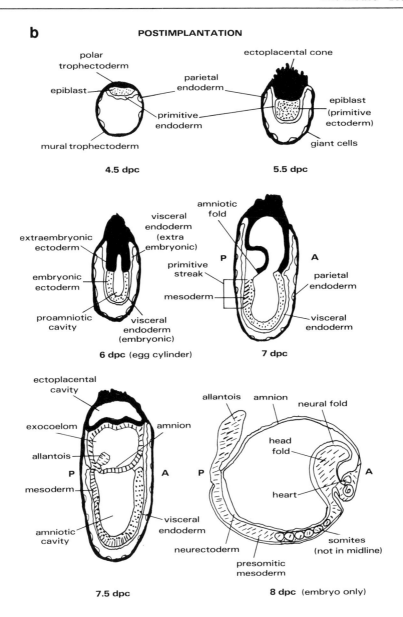

Figure 1. (a) Schematic representation of pre-implantation development. Description of the events can be found in the text, and an excellent series of photographs of the corresponding stages in ref. 2. Time of development is given in hours *post coitum* (p.c.). (b) Post-implantation development. Details are in the text. The time of development is given in days *post coitum* (dpc.). The anterior and posterior ends of the embryo, once the axis has formed, are indicated by A and P.

exocoelom. Below the exocoelom, separated by the amnion, is the amniotic cavity around which is the embryo proper. The chorion, derived from extraembryonic ectoderm in the fold, lies above the exocoelom, and defines the ectoplacental cavity. Other mesodermal cells form the allantois, a structure which grows up into the exocoelom where it eventually fuses with the chorion (from the ectoplacental cone) to give rise to the placenta and umbilicus.

In the embryo proper, however, cells recruited from the ectoderm move through an invagination, the primitive streak, to form an intermediate mesoderm layer between the endoderm and primitive ectoderm. Mesoderm formation extends anteriorly, with ectoderm cells moving through the primitive streak along the embryonic axis.

2.4 Primary induction

Mesoderm at the midline becomes the notochord. The mesoderm induces the ectoderm above it to become neurectoderm, the precursor of the central nervous system (CNS) and the neural crest. The rest of the encompassing ectoderm goes on to become epidermis.

The mesoderm sheet is divided into three parts either side of the midline, the presomitic mesoderm, the nephrogenic mesoderm and the lateral plate. Closest to the notochord, in the region of the trunk and tail, segmental structures, the somites, form (7,8). These groups of cells condense from the strips of presomitic mesoderm in a wave moving anterior to posterior (opposite to the direction of mesoderm formation). There is debate as to whether an underlying, determined, segmental pattern is present in the presomitic mesoderm (somitomeres). Certainly some workers report organized groups of cells in this region using scanning electron microscopy (9). Furthermore, unsegmented mesoderm will form somites in culture, away from any external inductive influences (10). Somites subsequently differentiate into three parts; sclerotome which gives rise to vertebrae and ribs, dermatome which forms the dermis and myotome from which body wall and limb muscles develop.

Outside the somites, another strip of mesoderm forms the nephrogenic organs; the pronephros (which also is segmented, at least transiently), the mesonephros and the precursor of the kidney itself, the metanephros. Most lateral of all, the lateral plate mesoderm, gives rise to the limb bud mesenchyme.

The third germ layer, the definitive endoderm, also arises from the embryonic ectoderm, probably at the anterior end, and forms the gut endoderm, the liver and part of the lungs.

It is the hope of studies outlined in this chapter to explain these events of early development, prior to organogenesis.

3. The molecular approach

One side of the developmental coin can be viewed by looking directly at gene expression in the embryo. This necessitates examining known gene products or simply looking for changes in unidentified proteins or transcripts. Enzyme microassays (11), SDS – PAGE of labeled proteins (12) and immunological techniques have been used to characterize proteins in pre- and early post-implantation embryos.

A pre-implantation mouse embryo contains less than 10 pg of poly(A)$^+$ RNA (13). Because of limited sensitivity, standard blotting techniques can obviously only be used to examine the most abundant transcripts, even if large numbers of early embryos are pooled. Newly developed techniques, based on the polymerase chain reaction (PCR) (14), use specific oligonucleotides to amplify selected transcripts of extremely low abundance or from extremely small amounts of starting material (15,16) and permit analysis even of rare mRNAs in single pre-implantation embryos (see Section 3.3 for examples).

Techniques have been developed which use carrier RNAs to enable cDNA libraries to be made from small amounts of embryonic mRNA (17,18). Using hybridization to these clones it is possible to quantitate specific mRNA abundance at pre-implantation stages (18). The localization of transcripts within the embryo can now be determined by *in situ* hybridization and examples are given below. This technique is currently limited in that rare transcripts are not detected but it may be possible to use some form of PCR *in situ* to amplify very low signals.

3.1 Gene expression in early mouse development

3.1.1 Maternal versus zygotic gene expression

One-cell mouse embryos treated with inhibitors of protein synthesis do not cleave, indicating that *de novo* translation is necessary for development to proceed. On the other hand, inhibiting zygotic transcription with α-amanitin allows the first cleavage to proceed, but the embryos arrest at the two-cell stage. Maternal mRNA is therefore exclusively used only briefly during development (19,20). On the other hand zygotic transcription in humans does not begin until the four- to eight-cell stage (21) and in sheep until the 16-cell stage (22). Degradation of maternal mRNA occurs concomitantly with the onset of zygotic transcription (20).

Among labeled proteins identified by gel electrophoresis at very early stages are three complexes of phosphoproteins, made from maternal mRNA, which undergo cell cycle-dependent phosphorylation (23). The first abundant products of zygotic expression are the heat shock proteins, Hsp68 and Hsp70, which appear early in the two-cell embryo (24).

3.1.2 Specific protein expression

Expression of a number of proteins has been characterized in pre-implantation embryos using immunological or enzymatic assays. A selection of these are shown in *Table 1*, and more can be found in Hogan *et al.* (1). Most proteins examined are candidates for functions expected in the embryo.

As tissues differentiate the proteins made in the embryo would be expected to become more diverse. However, when Wilkinson *et al.* (34) examined two-dimensional PAGE patterns of ^{35}S-labeled proteins in mouse embryos at 8 and at 10 dpc, between which times the somites begin to be formed, the only clear difference they could detect was the accumulation of globins in the developing blood system. Many proteins, including ones which play key roles in development, must be below the level of detection by these methods.

A lot of attention has been paid to protein and mRNA in the parietal and visceral endoderms, for which a model system for *in vitro* differentiation is provided by F9 embryonic carcinoma cells (35). Parietal endoderm cells have been shown to express, among others, type IV collagen, laminin and vimentin (35) which form Reichert's membrane. Visceral endoderm cells make a range of proteins including α-fetoprotein, transferrin, α_1-antitrypsin and others typical of embryonic liver (35,36).

3.1.3 Expression of mRNA in pre-implantation embryos

Technical limitations imposed by paucity of material has meant that little has been done with nucleic acid probes on pre-implantation embryos. Most work has concentrated on the abundant transcripts of repeated DNA. One example is the study of transcripts of intracisternal A-particles (IAPs), retroviral-like particles visible in cleavage-stage embryonic cells (37). They contain RNA transcribed from about 1000 retroposons in the genome (38). Probing RNA dot-blots with cloned IAP sequences show that IAP transcripts decrease after oogenesis, but subsequently increase from the two-cell to blastocyst stages (39).

Vasseur *et al.* (40) have used *in situ* hybridization to examine the accumulation of transcripts from another repeated DNA family, B2. They found B2 RNA at low levels in the nucleus of pronuclear stage embryos. The amounts increased during early cleavage stages. By late blastocyst the transcripts are localized primarily in the ICM, and by 7.5 dpc they are in the mesoderm and ectoderm, but not the visceral endoderm. Transcripts are mostly asymmetric, suggesting that the polymerase III promoters associated with the B2 elements are being used. Some opposite strand transcripts are seen, but these are always nuclear-localized, consistent with these being co-transcribed in pre-mRNA molecules and subsequently spliced out.

Taylor and Piko (18) also examined the transcription of the repeated

Table 1. Some proteins expressed in pre-implantation mouse embryos[a]

Protein	Stage/tissue	Detection method	Function?	References
Several phosphoproteins	One-cell	SDS–PAGE	Maternally encoded, cell cycle-dependent, phosphorylation	23
Hsp68, 70	Early two-cell	SDS–PAGE	Zygotically encoded heat shock proteins	24
Lamin A	Egg	Abs[b]	Nuclear structural proteins	25
Lamin B	Eight-cell/blastocyst			
β-1,4-galactotransferase	Compaction	Abs[b]	Cell surface, cell adhesion?	26
Na/K ATPase	Blastocyst (blastocoel inner face)	Abs[b]	Fluid accumulation in blastocoel cavity	27
Spectrin	Cleavage onwards	Abs[b]	Cell shape, calmodulin binding	28,29
Laminin	Morula	Abs[b]	Extracellular matrix	30
Uvomorulin (cadhedrin)	Compaction	Abs[b]	Ca-dependent cell adhesion	31,32
Gap-junction proteins	Compaction	Abs[b]	Cell communication	33

[a]This table lists only a selection of proteins expressed during pre-implantation mouse development. Others are listed in ref. 1.
[b]Abs indicates immunological detection using specific antibodies by immunohistochemistry or by Western blotting.

B2 and B1 families using slot-blots of cloned sequences probed with ^{32}P-labeled cDNA made from RNA of different embryonic stages. Absolute levels of B2 transcripts show an increase of about 60-fold between the two-cell and blastocyst stages. B1 RNA is present at a 10-fold higher level in two-cells, but only increases so that B1 and B2 abundance is the same in blastocyst. The significance of transcription of these repeated DNAs is not known, and no function has been ascribed to them. It is possible that they are simply a reflection of overall transcription occurring during development.

Taylor and Piko (18) also cloned 63 other, random, cDNAs from two-cell embryos, and examined the transcript levels of 37 of these. Most show an increase from the two-cell stage onwards of up to 100-fold, as did clones of β- and γ-actin and IAP. Twenty clones detected no transcripts in the unfertilized eggs or the one-cell embryo. The specific activation of these genes represents the first differentiative event of embryogenesis.

3.2 Oncogenes

Oncogenes are normal cellular genes which have become modified so that their expression level or pattern is altered, or their coding sequence changed to alter the function of the encoded protein and, as a result, have gained the function of at least partially bestowing oncogenic potential to the cell. The genes have either been incorporated from host genomes into oncogenic retroviruses or are modified endogenous genes. Given that oncogenes can modify the growth characteristics of cells, it is reasonable to assume that their progenitors, the proto-oncogenes, play a role in growth, and possibly development, of the embryo [reviewed by Adamson (41)].

It is therefore not surprising to find expression of these genes during development. The finding of differential expression, either spatially or temporally, is more interesting but it is difficult to propose, on these bases, a role for the products. Nor does the oncogenic potential of the genes necessarily indicate their normal function. Overexpression of the *int-1* and *int-2* oncogenes apparently results in mammary tumors in the mouse, for example, although the proto-oncogenes are not normally expressed in these cells, rather in the developing nervous system and migrating mesoderm cells (42,43). *Table 2* summarizes what is known about the expression of proto-oncogenes during embryogenesis.

The details of expression of proto-oncogenes are mostly too scant to allow one to postulate their role in development. In a few cases where the nature of the proto-oncogene has been identified a developmental role is possible, and no doubt others will be shown also to have a developmental and not just a cell cycle- or growth-related function. Ultimately loss-of-function mutations made in the genes might help to elucidate a role.

Table 2. Expression of proto-oncogenes during mouse development

Proto-oncogene	Function	Expression during embryogenesis	References
c-src	Protein kinase	Enriched in neural tissue	41
c-mos	Protein kinase	Primarily in testes and oocytes	
		Maternally encoded mRNA	44
c-abl	Protein kinase	Disappears at fertilization	45
Ha-ras		Very low	
Ki-ras	G-proteins	Widespread, no differential regulation	41,45,46
N-ras			
c-fos	Nuclear protein	Present in early embryos	45
		Absent later from embryo proper	
		High in extraembryonic tissue	
c-myc	Nuclear proteins (limited similarity to *Drosophila*	Widespread transcription, declines after birth	47,48,49
L-myc	*achaete scute* developmental genes)		
N-myc	*N-myc* declines after 11.5 dpc		
int-1	Membrane or secreted protein (homology to	Neuroectoderm-restricted at 8.5 dpc	42,50
	Drosophila segment polarity gene *wingless*)	Persists in localized regions of CNS	
int-2	Fibroblast growth factor	See Section 3.3.5	43,51
c-sis	PDGB-B	See Section 3.3.2	16,52
c-erb-B	EGF receptor	See Section 3.3.1	53,54

3.3 Growth factors

The list of polypeptide growth factors continues to lengthen. A full review of the field, even solely with respect to development, is beyond the scope of this chapter, but such a review has recently been published (52). Factors which regulate cell proliferation clearly can play a role in regulating development. There is both genetic and biochemical evidence that growth factors or related molecules can also regulate differentiation or morphogenesis. Two developmental loci of *Drosophila, decapentaplegic* and *Notch,* encode proteins with homology to transforming growth factor-β (TGFβ) and epithelial growth factor (EGF), respectively (55,56). In *Caenorhabditis elegans* the homeotic gene, *lin-12* also encodes a protein with EGF-like domains (57). Biochemical evidence comes from *Xenopus,* in which the induction of ectoderm to form mesoderm can be carried out *in vitro* by addition of members of the fibroblast growth factor (FGF) family (58) and which is enhanced by TGFβ (59) (see also Chapter 4). Furthermore, early *Xenopus* embryos naturally contain both FGF- (59) and TGFβ-related mRNAs. The latter is encoded by an mRNA (Vg1) which is maternally encoded and localized to the vegetal part of the egg (60). The developmental distribution of expression of the five families of growth factors in mouse is outlined below.

3.3.1 Epithelial growth factors

EGF is not produced by the mouse embryo until mid-gestation or later. However, the EGF receptor, the product of the *c-erb-B* proto-oncogene (53), can be shown to present (by binding of labeled EGF) on the surface of trophoblasts growing out *in vitro* from explanted blastocysts (54). This receptor is bound in early embryogenesis by another EGF family member, TGFα (61). Recently, Rapollee *et al.* have used PCR amplification of cDNA to demonstrate that the TGFα mRNA is in fact present as a maternal mRNA in fertilized eggs. Only very low levels of the message persist but the amounts increase before the blastocyst stage (16).

3.3.2 Platelet-derived growth factors

There are three forms of PDGF, homodimers of A or B chains, and the heterodimeric form. PDGF-B is the product of the *c-sis* proto-oncogene. It is reported that both A and B mRNAs are present at gastrulation (52), and that the PDGF-A mRNA, at least, is detectable by PCR-amplification as a maternal message which disappears at the two-cell stage to reappear as a zygotic transcript (16).

3.3.3 Insulin-like growth factors

It is thought that insulin-like growth factor II (IGF-II) is the predominant IGF during embryogenesis. Antibodies directed against IGF-II specifically precipitate a protein from mouse yolk sac mesoderm and amnion, but not from visceral or parietal endoderm (62). Later human embryos express both IGF-I and II, as shown by Northern blotting and *in situ* hybridization, largely in tissues of mesodermal origin, and in a way consistent with these cells exerting an influence on neighboring cells (63).

3.3.4 Transforming growth factors

The synergistic effect of TGFβ with FGF on *Xenopus* mesoderm induction was noted above. TGFβ activity has been detected in late-gestation mouse embryos and TGFβ1 mRNA has been identified as a zygotic transcript at the two-cell stage (16). Another member of the TGF family, Mullerian inhibiting substance (MIS), has undisputed developmental activity. MIS acts later in mammalian embryogenesis to cause regression of the Mullerian ducts during male gonadal development, and activation of its synthesis by the developing test may be a primary effect of the testes-determining gene on the Y-chromosome.

3.3.5 Fibroblast growth factors

The FGF family includes basic and acidic FGF and a growth factor isolated from embryonic carcinoma cells, embryonic carcinoma-derived growth factor (ECDGF). The production of such a factor by EC cells (64) strongly suggests that it has an as yet unknown function during early embryogenesis. Other sequences have been identified as having homology to FGF, including the human oncogene *hst* (65) and the mouse *int-2* oncogene (51). Like *int-1* (to which it has no sequence similarity) *int-2* expression is activated in some mouse mammary tumors. It is otherwise embryo-specific in its mRNA expression (43). At mid-gestation it is expressed in neuroepithelia, in the endoderm of the pharyngeal pouches and extraembryonically in the parietal endoderm. Early in embryogenesis, at gastrulation, it is detected by *in situ* hybridization in the mesoderm cells as they migrate through the primitive streak.

The preceding catalog gives little indication of what role, if any, growth factors play in determining or regulating mouse early development, although there is sufficient evidence to believe that they must. Specific localization of growth factors early post-implantation must be determined, and the target cells identified by virtue of their receptors. In addition, it is clear that combinations of growth factors will modulate the effects of each single factor, and so any simple interpretation may be hazardous.

3.4 Homeobox genes

The homeobox was originally characterized as a short, 183 bp sequence, encoding a 61 amino acid domain, shared between several homeotic genes of *Drosophila* (66,67) (see Chapter 2). The initial homeobox-containing genes discovered were confined to the two major homeotic gene clusters, the *Antennapedia* and *Bithorax* complexes (*ANTP-C* and *BX-C*). Since then, other, more diverged, homeobox containing genes have been identified at other loci. The genes of *ANTP-C*, *Deformed* (*Dfd*), *Sex combs reduced* (*Scr*), *fushi tarazu* (*ftz*), and *Antennapedia* (*Antp*); and the *BX-C*; *Ultrabithorax* (*Ubx*), *iab-2* and *iab-7* form a distinct family, where most members have more than 70% amino acid identity in the homeo-domain. The segment polarity gene *engrailed* (*en*) and a linked gene, *invected*, form another family, closely related to each other but with less than 50% amino acid identify to the homeodomain of the *Antp* family. The *paired* group of segmentation genes consists of at least three genes, with homeoboxes considerably diverged from both *Antp* and *en* (68). In addition each has a second non-homeo'box' (see Section 3.5). Other genes with homeoboxes have no apparent segmentation function. The maternal-effect polarity gene, *bicoid* (*bcd*), for example, has about 40% homeobox identity to *Antp* (69), whereas *rough*, an eye-pattern gene (with additional effects in the nervous system), has 57% identity (70).

Sequence comparison has found elements recognizable as homeo-domains in evolutionarily very distant species. The yeast mating-type genes, *mat-a1* and *mat-α2* (71) and another yeast regulatory gene, *PHO2* (72), have recognizable homeodomains but with as little as 28% amino acid identity to the *Antp* box. Likewise, a *C.elegans* developmental gene, *mec-3* has a protein domain which has 42% amino acid identity to the *paired* homeobox, but only 25% (15/61) with *Antp* (72). The range of sequences which have been termed homeoboxes is much wider than was at first anticipated. It remains to be seen whether sequence similarities between the homeodomains and family groupings can be equated with functional similarities within or across species.

The evidence is now good that the homeobox encodes a DNA-binding protein domain. The *Drosophila* proteins are localized in the nucleus (67), and there is some evidence for sequence-specific binding (74,75). It is proposed that part of the sequence is a helix-turn-helix motif, analogous to the structure of lambda repressor (71,76). If that is so, the second helix would form the recognition helix for specific DNA binding. In fact it is this region of the domain which is the most conserved, even between the most widely diverged sequences.

3.4.1 Mouse Hox gene structure

When probes from the *Antp* family are used to examine vertebrate DNA, a number of fragments clearly hybridize (77). Twenty or more mouse homeobox (*Hox*) genes have now been isolated on this basis, almost all forming an *Antp*-like family. These fall into four clusters of linked genes. Mouse genes more closely related to the *en* sequences have been isolated, as have other more distant genes (see below). It is probable that other diverged homeobox sequences will be identified.

The four clusters of *Antp*-like homeobox genes are shown in *Figure 2*. *Hox-1* and *Hox-2* each contain at least seven members. Sequence

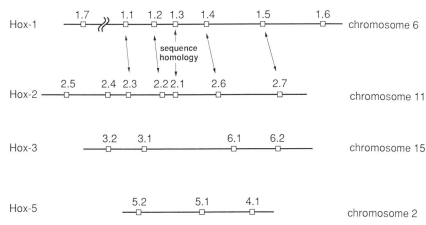

Figure 2. The major mouse homeobox (*Hox*) clusters. Other *Hox* genes are found singly on other chromosomes. The spacing of the *Hox-1* and *Hox-2* clusters is approximately to scale. The alignments between genes as indicated by the arrows is based on sequence similarity, and probably shows an ancestral relationship between the clusters. Similar alignments with the other clusters may be possible but are not shown. See refs 78–84.

relationships between complete cDNA sequences, or the homeodomains alone (78,79) in the *Hox-1* and *-2* clusters, are indicated by the alignment shown and suggests they have evolved by duplication of a single ancestral cluster. It is likely that more *Hox* genes will be found in all four clusters.

In general, each gene consists of two exons, the second of which does not encode much more than the homeodomain. In every case the homeobox is preceded, a few codons upstream, by the intron, which is usually 1–2 kb. Some cDNAs representing alternatively spliced transcripts have been seen from mouse *Hox-1.6* (84,85) and from a *Xenopus* gene (86). These could encode variant proteins, in some cases without the homeodomain. In human placenta, evidence has been found for transcripts running down much of the *Hox-3* cluster, and splicing together of exons from different 'genes' (87). It remains to be seen if any of these alternately spliced products are functional.

Where *Hox* cDNA sequences have been examined, a sequence related to the hexapeptide Ile—(Phe/Tyr)—Pro—Trp—Met—Arg is encoded a few amino acids upstream of the box but always on a different exon (89). This hexapeptide is also seen in many *Drosophila* homeobox genes, and may be a 'hinge' region, separating the homeodomain from the rest of the molecule. Comparisons of the protein sequences encoded by homologous *Hox* genes [e.g. *Hox-1.3* and *Hox-2.1* (89)] show that conservation is concentrated at the ends of the molecule, around the homeodomain and around the N terminus, with virtually no amino acid similarity in the middle part. It is tempting to ascribe a function to this structure, in which the C terminus interacts with DNA and the other end perhaps interacts with other proteins.

There are two homeobox genes in the mouse more closely related to the *Drosophila en* gene than to the major *Hox* loci (90). The similarity extends beyond the homeobox for a total of 95 amino acids. These two genes, *En-1* and *En-2*, are unlinked and appear to have evolved by independent duplication in both species (R.Hill, personal communication).

Recently, two other loci have been described, *Hox-7* and *Hox-8* (91). They appear, so far, to be single genes, mapping on different chromosomes (R.Hill, personal communication). They are substantially diverged from the *Antp* family of genes (about 45% identity), but are much closer to *msh* (muscle-specific homeobox) of *Drosophila* (also known as 99B) (92).

3.4.2 Hox gene expression

The rationale for this biochemical approach to the isolation of mouse genes is that the conservation of homeobox mutations reflects selection for a function maintained between *Drosophila* and mouse. The fact that evolution of the major *Antp*-related *Hox* clusters has been constrained much more than other homeobox sequences lends hope that the conservation is for more than simply DNA binding. No mouse *Hox* gene has been shown to be allelic to any mapped mutation. The approach taken to analyze the genes has been initially a descriptive one, localizing their expression during embryogenesis. An excellent review has recently been published (93).

Parallels with *Drosophila* would lead us to expect the crucial time of *Hox* expression in mouse to be when key early developmental decisions are made (i.e. gastrulation to early somite stages, ~7.5 – 8.5 dpc). Indeed, expression of *Hox-1.3* and *-2.1* has been detected 7.5 dpc by RNA analysis (89,94), and of *Hox-1.5* and *-3.1* by *in situ* hybridization (95). The spatial distribution at early somite stages (~8 dpc) has been determined for mRNAs of *Hox-1.5, -2.1, -3.1* and *-6.1* by *in situ* hybridization (95 – 97). Transcripts are seen in both the embryonic ectoderm and mesoderm, plus the allantois. Each of the genes occupies a different domain along the anterior – posterior (A/P) axis of the embryo. *Figure 3a*, from the work of Gaunt (95), shows *in situ* hybridization of *Hox-1.5* and *-3.1* to adjacent sections of an 8 dpc embryo, demonstrating the expression of these two genes. *Figure 3b* is a schematic of the expression domains of four *Hox* genes. The ectodermal and mesodermal domains largely coincide, although it is possible that limit of expression in the anterior mesoderm is, in some cases, slightly behind the limit of expression in ectoderm. This type of pattern is strikingly similar to the expression of *Drosophila* homeotic genes, where different members of the *ANTP-C* and *BX-C* are expressed in ectoderm and mesoderm in discrete but overlapping domains.

Other genes have usually only been analyzed later in development and almost all are expressed in the spinal cord or brain. The relative anterior

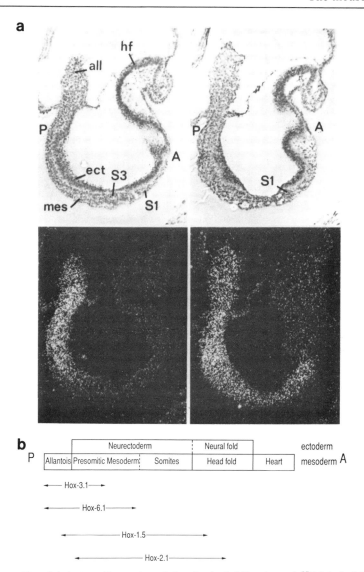

Figure 3. (a) Autoradiographs of *in situ* hybridization of ^{35}S-labeled RNA probes to sections of 8 dpc embryos. *Figure 1b* shows a schematic drawing of the same approximate stage. The left panels show hybridization to *Hox-3.1*, and the right panels, hybridization to *Hox-1.5*. The upper panels show phase-contrast optics, whereas the lower panels show dark field illumination to reveal the silver grains. A, anterior; P, posterior; hf, headfold; all, allantois; mes, mesoderm; ect, ectoderm; S1, S3, somites 1 and 3. The original photographs were kindly supplied by Dr S.Gaunt and the figure is taken from ref. 93. (b) Diagrammatic representation of the regions of 8 – 8.5 dpc embryos which express four *Hox* genes, as determined by *in situ* hybridization. The arrows indicate the approximate extent of silver grains observed along the A/P axis. The blocks above represent regions along this axis of ectoderm and mesoderm. The data are taken from refs 93, 95 – 97.

boundaries of expression in the CNS of *Hox-5, -2.1* and *-3.1* at 12.5 dpc do seem to correspond to their relative positions at earlier stages and it may be possible to extrapolate other expression patterns back to somite stages.

Expression is also seen to persist in mesodermally derived tissues; lung, kidney, gut (79,89,93,94,96) and in somite derivatives (93,96 – 99). Again it is possible that this expression could be extrapolated back to the spatial restriction in the mesoderm at 8 dpc, in which case the expression of particular genes would be useful lineage markers, to trace the origins of these organs back to particular regions of the mesoderm layer. It is necessary to examine the temporal pattern of expression more carefully.

Where the domains of CNS expression and somite-derived expression can be readily compared in later embryos (e.g. *Hox-6.1*) (96), the anterior limit in the somite derivatives is considerably behind the boundary in the spinal cord. It is not known if this results from relative cell movement through development, or a real difference in domains of expression in the two germ layers. It may be significant that expression of *Ubx* in *Drosophila* embryos is in both mesoderm and ectoderm, but the limit of ectoderm expression is more anterior than in the mesoderm.

The *En-2* gene is expressed from early somite stages only in a very restricted region of otherwise homogeneous neurectoderm, anterior to all the *Hox* expression domains (100,101). Expression persists in this region of the CNS throughout embryogenesis and into the adult brain. *En-1* is expressed in exactly the same parts of the nervous system as *En-2*, but also in a range of other tissues including the posterior part of the hindbrain and the rest of the neural tube, the prevertebrae, the dermatome portion of the early somites and parts of the developing limb bud and facial mesenchyme (101).

To focus simply on a role in positional determination or segmentation for *Hox* genes may be an error. In *Drosophila*, genes of the *ANTP-C* and *BX-C* are expressed later in development in particular cells of the CNS. Doe *et al.* (102) were able, by manipulating *cis*-acting elements of the *ftz* gene, to permit early segmental expression whilst abolishing the later neuronal expression. Flies with only this *ftz* gene had altered neuronal fates. Perhaps the consistent expression of *Hox* genes in the mammalian CNS indicates a similar role.

No *Hox* gene expression has been detected earlier than 7.5 dpc in mouse. In *Xenopus* also most genes are expressed at gastrulation, but several are additionally expressed in oocytes (103,104) and this mRNA disappears after fertilization. A number of *Drosophila* homeobox genes also are expressed maternally. PCR-amplication of mRNA will be needed to see if *Hox* transcripts are present in mouse eggs and cleavage-stage embryos.

If the mammalian *Hox* genes are mediators of positional information in the developing embryo, many questions remain. What regulates the

Hox genes in the mouse gastrula, that is what is the primary determinant of the domains of expression? To what other genes do the *Hox* gene products pass on positional information, that is what are the effective targets of the DNA binding function? Is the DNA binding effective with the *Hox* gene product alone, in combination with other *Hox* gene products or in combination with other proteins? Finally, the question of function of the *Hox* genes needs to be approached using mutagenesis (see Section 5).

3.5 Other mouse homologs to developmental genes

Sequence similarities between gene products are useful in formulating conclusions about the function of genes. The similarities between *Drosophila decapentaplegic*, the *Xenopus* Vg1 product and mammalian TGFβ have already been noted above (55,60). Similarly, screening by hybridization across phyla has produced notable success in identifying the *Hox* genes. Screening in reverse, using a mouse probe to identify the *Drosophila* homolog of the *int-1* gene, picked out the *wingless* locus (50). The isolation of more *Drosophila* genes has prompted further screening of mouse libraries to pick out mammalian homologs. The best hope for success must be when a particular sequence is found in several, otherwise dissimilar, *Drosophila* genes.

3.5.1 The paired box

The 'paired box' is found in a number of *Drosophila* segmentation genes (which also contain a homeobox) (68). The paired box encodes 126 amino acids, with 80–90% identity between the different genes, and a paired box probe identifies a family of genes in the mouse. One of these genes, named *Pax-1* (105), has a paired box with around 70% identity to the fly. The gene is not expressed during early development. It is first detected at 9 dpc in the sclerotome, a differentiated product of the somites, and continues to be expressed in the prevertebrae. It is probably expressed too late to be involved in primary segmentation, and any regulatory function it encodes must be acting on later events (see Section 4.1).

3.5.2 Zinc finger proteins

Krüppel and *hunchback* are *Drosophila* gap genes, which regulate development at an earlier stage than the segmentation genes (see Chapter 2). Both contain a domain consisting of a number of so-called 'zinc fingers' (106,107), a moiety first recognized in the *Xenopus* transcription factor TFIIIA (108) (see Chapter 4). The finger is a DNA-binding unit of about 28 amino acids containing two cysteines and two histidines which probably complex a zinc atom, together with a number of other conserved residues. Other DNA-binding or putative DNA-binding proteins also have finger domains (109). A probe from the finger domain of the *Krüppel* gene hybridizes to mouse DNA (110), and has been used to isolate a number of genomic and cDNA clones. The part of the domain most similar between all these clones is not the zinc-binding element but a presumptive 'inter-finger' region; which is not actually found in TFIIIA.

The expression of one mouse finger gene, *mKr2*, determined by Northern blotting and *in situ* hybridization is restricted mainly to the adult and developing nervous system from 10 dpc onwards (111). No expression is detected at 8.5 dpc, although cDNAs have been isolated from an 8.5 dpc library. In the absence of more information it is not possible to postulate a role for *mKr2* in early development, but the finger genes provide interesting candidates.

4. The genetic approach

A complete picture of gene function requires that the biological effects of abnormalities in expression are known. For most of the genes described in Section 3, no such abnormalities exist (but see Section 6 for a description of how they may be generated). An alternative approach is to isolate genes for which there are mutations. This can be done in two ways; either genes identified in mouse stocks or induced at particular loci can be cloned, or mutations can be made by insertion of DNA into the genome at random. Mutant genes identified by the latter route should be tagged by the presence of the inserted DNA and can be cloned using the insert as a probe.

It must be borne in mind that not all genes which, when mutated, cause embryonic death are developmental genes in a strict sense, indeed the majority are not. Any cell-lethal will kill the embryo, probably after a few cell divisions have proceeded and the maternal contribution exhausted. Mutations in genes required in certain cell types will kill embryos at later, specific times.

Genetic abnormalities seen in viable mice can often be traced back to defects in early embryogenesis. The basis for that defect may, in a sense, be trivial. A defect or deficiency in an enzyme important for cell proliferation might slow down cell growth at a critical time (such as mesoderm formation or somitogenesis) and lead to abnormalities. Treatments which cause cell death in the embryo will often result in malformed offspring, particularly with defects of the axial skeleton. For example, Snow and colleagues (112,113) have shown that mitomycin C treatment *in vivo* can kill up to half the cells in an embryo, but if it is then removed, the embryo (now with only one-tenth the number of cells compared to an untreated control) will recover, undergo restorative growth to 'catch-up' with the developmental timetable, and be born no smaller than untreated mice. The restorative growth phase, however, appears to induce specific abnormalities of the vertebrae and ribs, which are very like those due to particular genetic mutations. In other species, mutations in 'housekeeping' genes can produce developmental defects. For example, the *rudimentary* locus of *Drosophila* encodes an enzyme involved in pyrimidine synthesis, but mutations in the gene cause specific defects in wing development (114).

4.1 Genetics of early mouse development

It is virtually impossible to identify a novel mutation in mouse which is only a recessive lethal. For this reason, most homozygous lethals which have been studied also have a dominant, visible phenotype. Some notable exceptions are at the coat color loci and in the *t*-complex where particular regions of the genome are followed using closely linked markers after mutagenesis (see below). It is important to bear in mind that there is not necessarily a relationship between the heterozygous effect and the cause of homozygous lethality. A mutation, especially a deletion, may affect several genes. The dominant phenotype can arise from haplo-insufficiency or altered expression one of the genes, whereas embryonic death might arise because of homozygous loss of function of a different one. *Lethal yellow* is an example of this type of mutation (see Section 4.2).

Genes which cause early embryonic death are reviewed by McLaren (115). I will discuss only a few of these, for which a molecular approach seems feasible.

Oligosyndactyly (*Os*) has a dominant effect, causing fused digits. *Os* homozygotes die very early in development, about the blastocyst stage (116,117), apparently due to arrest of the cell cycle at metaphase after 5 – 6 divisions. Presumably the wild-type *Os* gene encodes a protein required for mitosis, which is not required until later divisions, either because another embryonic gene supplies it until then, or because a maternal protein persists and is used up to that point. Chimeras between wild-type and homozygous embryos do not rescue the mitotic arrest (118). The mutants can be transiently rescued by the injection into each blastomere of the two-cell embryos of mRNA from embryonal stem cells (118). This procedure, although difficult, might provide an assay for testing the ability of candidate genes to rescue, or even for the isolation of the gene.

Molecular access to other mutants will generally be gained in the end through closely linked DNA markers from which long-range mapping and cloning techniques will be used, or by the identification of candidate genes, which will be tested for genetic localization followed by rescue in transgenic animals.

Two loci for which close DNA markers are being generated are *Fused* (*Fu*) and *Brachyury* (*T*). These both map to chromosome 17, in the *t*-complex, although *Fu* and *T* arose on wild-type (non-*t*) chromosomes. Mice heterozygous for three *Fu* mutant alleles (*Fused, Kink* and *Knobbly*) have tail defects (115,119). *Kink* and *Knobbly* are particularly interesting homozygous lethals. (*Knobbly* is not a fully penetrant lethal, about 10%

of homozygotes survive.) They have abnormalities of the embryonic ectoderm at 7 dpc which is possibly defective in the setting up of the embryonic axis; duplications of the neural tube and other structures are seen. There is also a high incidence of normal and abnormal twin and even triplet embryos (multiple embryos in the same decidua) seen in crosses between carriers of these mutations. An elegant genetic analysis (119) showed that the dominant characteristics are gain-of-function mutations, rather than haplo-insufficiencies. The isolation of molecular probes around the t-complex (see Section 4.3) should ultimately lead to the isolation of the *Fused* locus and elucidation of the nature of the dominant and recessive phenotypes.

Sequences very close to or within the *Brachyury* locus have already been isolated (B.Hermann, personal communication). This mutation causes a short tail when heterozygous with another wild-type chromosome 17, but produces tail-less mice when opposite a chromosome carrying the t-complex (Section 4.3) through an interaction with a mutation, *tct*, in the complex. Embryos homozygous for *Brachyury* die at mid-gestation, although they become abnormal much earlier, at 8 dpc, when the primitive streak and notochord fail to develop properly. Several other mutant alleles of *Brachyury* have been studied (115,120).

Other mutations, not lethal during development, nevertheless have clear morphological effects which can be traced back to abnormalities early in embryogenesis. Several of these have effects on segmentation (7,8). One of the most severe is *pudgy*, a recessive mutation on chromosome 7. Homozygotes have gross abnormalities in the axial skeleton, including fused and abnormal ribs and vertebrae. The defect can be identified at the somite stage, when the presomitic mesoderm forms correctly, but is not divided adequately into somites.

Two segmentation mutants map close to genes with homology to *Drosophila* segmentation genes (see Sections 3.4 and 3.5). *Hox-5.1* maps on chromosome 2 close to *rachiterata* (83) and *Pax-1* maps close to *undulated*, also on chromosome 2. Allelism between the mutant loci and the cloned genes has not yet been proven but there is an amino acid difference between the normal and *undulated Pax-1* sequences (121).

4.2 The coat-color gene complexes

The most studied genetic system in the mouse is that involved in the development of hair pigmentation. There are many loci which affect coat coloration, some of which have several alleles. Some alleles of certain loci have other effects on embryonic development. Furthermore, six coat-color genes, *non-agouti, brown, chinchilla, dilute, pink-eyed dilution* and *piebald*, are used in the specific locus mutation test in which wild-type mice are mutagenized and crossed with mice homozygous for these recessive mutations (plus the morphological mutation *short ear*) to reveal new mutations at the loci. Many new alleles of the mutations have been

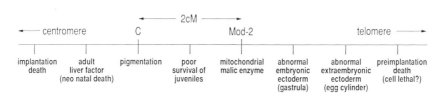

Figure 4. Genetic deletion map around the *albino* (*c*) locus. *Mod-2* and *c* have been mapped genetically relative to each other and to the centromere and telomere by standard crosses. All other loci marked are deduced from phenotypes of mice homozygous for various deletions, and mapped by complementation. This map is based on data in refs 123 – 125.

identified. Some, particularly those induced by radiation, are also homozygous lethal, presumably due to deletion of neighboring genes which are necessary for development.

The best studied deletion series is around the *albino* (*c*) locus where the *chinchilla* tester gene is located. Mutagenized progeny (3.6 million) tested gave rise to 199 new mutations at *c*, of which 39 have a recessive lethal or subvital phenotype (122). Some of the lethalities act late in development, even after birth; others act early. By means of a considerable amount of work, 34 have been tested in pairwise combinations for complementation of the lethal phenotype, and the approximate stage of death of all the lethalities examined. A deletion map of genes around *c* has been derived (123), and recently been refined to separate two gastrulation lethal complementation groups (124). *Figure 4* is a schematic of the genetic map. Only a subset of genes are identified in this type of deletion map as any deletion beyond a specific stage lethal will only reveal lethalities acting earlier in development. Furthermore, two genes with the same stage of lethality will not be separated by the analysis. The genetic map can be refined and all the genes in the region identified by chemical mutagenesis, selecting for visible or lethal mutations spanned by a large deletion (E.Rinchik, personal communication).

Embryos with homozygous deletions extending on the distal side of *c* to the 'embryonic ectoderm' function have normal extraembryonic tissues as late as 8.5 dpc. Embryonic mesoderm is also produced (as is the allantois), but the embryonic ectoderm is abnormal (124). Larger deletions in the same direction result in homozygous embryos with underdeveloped and disorganized extraembryonic ectoderm at the egg cylinder stage (124). An even larger deletion in the same direction removes a gene required very early in development, and the homozygous embryos die at the two- to four-cell stage (125). This could well be a cell-lethal, as the embryos arrest at about the time when zygotic transcripts are required. DNA probes from the region of the *c*-locus have been isolated (E.Rinchik, personal communication), including tyrosinase, the product of the pigmentation gene itself (126,127). The extensive set of nested deletions, plus the new chemically induced mutations will greatly aid molecular analysis of the region.

Exactly analogous work has been done around the *dilute* (*d*) locus, and the closely linked *short-ear* (*se*) gene. Five genes required for pre-natal survival of the embryo have been defined by deletions around *d-se* (128,129). Morphological descriptions of the lethalities are still lacking, although molecular analysis of the region is extensive (130). Recessive lethals have been induced also around the *brown* and *pink-eyed dilution* loci. These have been little characterized, but new molecular probes (131) will aid future work.

The *non-agouti* (*a*) locus has very interesting genetic features. Several alleles are homozygous lethals. One of these, *Lethal yellow* (*A^y*), was the first recessive lethal gene to be identified in the mouse (132). In fact *A^y* acts as a pseudoallele of *a*, in that wild-type offspring can arise in crosses between mice doubly heterozygous for both *A^y* and other mutations at *a* (133,134). The simplest model is that there is a mutation which is recessive lethal and which also acts at a genetically separable distance to confer the dominant mutation on the wild-type *a* gene. The lethal effect is exerted early in development. Homozygous *A^yA^y* embryos are normal-looking blastocysts but die around implantation. Inner cell mass cells from these embryos, transferred to wild-type blastocysts, survive in chimeras, but normal ICMs will not rescue mutant blastocysts (135). The defect is therefore not cell-lethal, and is localized to the trophoblast lineage.

Four other *non-agouti* alleles are homozygous lethal. Crosses between them show they fall into two complementation groups, one also containing *A^y* (134). Three of the four are radiation-induced, and may be deletions, some of which encompass *A^y*. A member of the second group, *lethal light-bellied non-agouti* (*a^x*) has a lethal effect later (136). Homozygotes are visibly abnormal as blastocysts, but they do implant, and a range of abnormalities are seen; different dying embryos having different cell types that are either missing or abnormal. Recently DNA has been cloned from the region (134,137), opening up the locus for molecular study.

4.3   The *t*-complex

The *t*-complex is a segment of chromosome 17 of approximately 30 megabases with a unique set of properties (138,139). Certain mice carry a chromosome which has two inversions within the region with respect to wild-type (*Figure 5*). Recombinations between the *t* (inverted) chromosome and wild-type is therefore suppressed and markers along the *t*-chromosome are inherited as a unit, the *t*-haplotype. The *t*-haplotype has evolved a mechanism for invading the population. It shows male transmission ratio distortion. This is a phenomenon mediated by a number of distorter genes acting on a responder gene, all within the haplotype. It results in the *t*-chromosome being transmitted through the sperm of

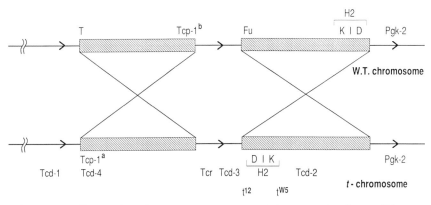

Figure 5. The region of chromosome 17 around the *t*-complex from wild-type and from *t*-bearing mice. Recombination between wild-type and *t*-chromosomes is suppressed by the two large intrachromosomal inversions, depicted by the boxed segments. The chevrons illustrate the orientation and point towards the telomere in the wild-type chromosome. Transmission ratio distortion is mediated through a receptor locus (*Tcr*) acted on by at least four distorters (*Tcd-1* through *4*), all borne by *t*-chromosomes. The developmental mutations *Brachyury* (*T*) and *Fused* (*Fu*) map to the region, as does the major histocompatibility complex, *H-2*. *Tcp-1* is the gene for the *T-complex polypeptide-1*, which has different electrophoretic properties when isolated from wild-type or *t*-bearing mice. The testes-specific phosphoglycerate kinase gene (*Pgk-2*) maps close to the distal end of the complex. Two *t*-haplotype lethals, t^{12} and t^{w5} which map very close to *H-2* are shown on the map. The data are taken from refs 139 – 141.

heterozygotes up to 100 times more frequently than its wild-type partner. However, the distorter genes cause male sterility when homozygous. A large number of sterile males in the population is evolutionarily disadvantageous, and there is selective pressure in favor of recessive lethals occurring within the *t*-haplotype, to prevent these males being born. The recombination suppression means that the transmission ratio distorter genes and the lethals are not separated, and so the system equilibrates.

Many *t*-haplotypes have been isolated from laboratory and wild mice. The lethals on different *t*-haplotypes may complement, either identifying new lethal genes, or falling into one of the known groups. Most *t*-chromosomes carry only one lethal mutation, although some carry more. At least 16 complementation groups have been identified (115,120, 139,142).

When viewed in this light there is nothing special about *t*-lethals, they simply arose by chance within the haplotype, and bear no particular relationship to each other. Nevertheless, they have received extensive study, and the good genetics and molecular characterization of the region makes the genes prime candidates for cloning. More lethals which map to the region have been produced by chemical mutagenesis of wild-type chromosomes (143).

The effects of *t*-lethals on development have been reviewed (115,142).

Lethalities have been found in different homozygous t-haplotypes which include among others:

(i)   failure to develop from morula to blastocyst (t^{12});
(ii)   failure to implant (t^{w73});
(iii) failure of primitive ectoderm to differentiate to embryonic and extraembryonic components and to elongate the egg cylinder (t^0);
(iv)   failure to maintain the embryonic ectoderm after the egg cylinder stage (t^{w5});
(v)   a defect in the primitive streak, blocking cells forming mesoderm (t^{w18}).

The work on the embryology of the t-lethals justifies certain efforts to clone the genes. The cloning of many DNA markers throughout the complex means that there are probes close to some. The major histocompatibility complex, *H-2*, which has been extensively cloned is within the t-complex, at the end of one of the inversions. Two lethals, t^{w5} and t^{12}, map very close to or within *H-2*, and candiate genes have been identified (144; H.S.Shin, personal communication).

4.4 Insertional mutagenesis

The cloning of genes identified by classical genetics from their mutant alleles may be an extremely laborious process. There must be a trade-off between the difficulty in cloning the gene, and the knowledge gained once it has been isolated. As there is nothing particularly outstanding with respect to development about the mutants already known, and given the relative ease by which mutations due to insertion of foreign DNA can be cloned, it may be that the most effective way to study mutants which affect development is to make them by insertional mutagenesis.

DNA introduced into the genome to make transgenic animals for the study of gene expression may cause mutations (145). Often such animals are tested routinely for a homozygous effect of the transgene. Other experiments are geared specifically to cause insertional mutations, usually employing retroviruses to introduce DNA efficiently into embryos or embryonal stem cells.

Estimates of the frequency by which inserted DNA causes a mutation are around 5 – 15%. The majority of mice examined therefore have no homozygous phenotype. Those that do will most often be embryonic lethals, although some viable, morphological mutations have been found. Two such mutations are allelic to *limb-deformity* and *downless* (146; P.Overbeek, personal communication). Methods have been developed using embryonal stem cells which allow many insertions to be followed simultaneously (see Section 6). There are also strains of mice which, when crossed, allow mobilization *in vivo* of previously ineffective retroviruses (147). This may be a way to generate novel insertions with relative ease. Insertional mutations have only recently been studied, and therefore not a great deal is known about the embryology of the defects.

Retroviral infection of embryos has produced an insertion into the α1(I)collagen gene (the Mov-13 line) (148), which when homozygous causes embryonic death at mid-gestation probably because the major blood vessels rupture (149). Another retroviral insertion is recessive for lethality in the adult, apparently due to kidney failure (R.Jaenisch, personal communication). Recently a novel insertional mutant has been described, named *legless*, which has effects when homozygous on both hindlimb and brain development. These effects are first manifest in early development (150).

Two recessive retroviral insertions, MPV-20 and Mov-34, which cause early embryonic death have been reported. Mov-34 homozygotes die soon after implantation (151). The cloned DNA flanking the provirus insertion in Mov-34 detects a widely expressed transcript (145,151). Other early embryonic mutations have been produced by integration of injected (transgenic) DNA. Lacy and colleagues (152) have identified an insertional mutant (line 4) which arrests around the time of implantation.

Embryos homozygous for quite a number of transgenic integrations, in addition to Mov-34, will implant but die or become abnormal soon after. This might be indicative of a widespread activation of new gene expression around this time. Two such recessive mutations, HUGH/3 and HUGH/4, are caused by integration of a human growth hormone gene (153). Another is HB58, caused by the integration of a human globin gene. This mutation allows implantation and normal extraembryonic development, but results in the embryo undergoing abnormal neuralation without forming somites (F.Costantini, personal communication). Finally a mutation, cv4, due to insertion of an actin gene, has been shown to cause death before mid-gestation, but has not been further defined (154).

The rationale for defining these mutations is that the DNA flanking the insert will be from the mutant gene. Retroviruses insert cleanly into the genome and there is not usually any particular difficulty in cloning and analyzing the integration site. However, when DNA is introduced by microinjection into embryos, the DNA may integrate either as long concatamers, or interspersed with host DNA, and furthermore, may cause rearrangements of the host genome at the site of integration. Any one of the events may create problems in cloning the flanking DNA, and in analyzing the nature of the mutation. This is exemplified by HUGH/3 (153) in which there is a deletion at the site of multiple, tandem copies of the human transgenes interspersed with mouse DNA (155). The line 4 integrant (152, see also 145) has generated a long (more than 18 kb) deletion. If large deletions are produced, several genes could be disrupted and it could prove difficult to identify the one which causes the phenotype.

5. Transgenic mice

In the 10 years since the first transgenic mice were reported the technique has become a widely used method for analysis of gene expression (156).

Cloned genes microinjected into a pronucleus of the fertilized egg integrate into the genome with a frequency of up to 25%, and in most cases the 'transgene' is found in all cells of the resulting mouse, including the germ line. Transgenic animals can also be made by retroviral infection of embryos pre- or post-implantation (157,158). Much work with transgenics has been concerned with the identification of DNA sequences required to confer spatial and temporal specificity on gene expression (156). Ultimately this approach will be applied also to genes expressed early in development. The *Hox* genes, for example, are receiving this kind of attention. I will examine here other aspects of transgenic animals which tell us about early development. First, transgenes can be expressed inappropriately, by chance or by design. Cases where the expression has been directed to the wrong tissue, or a modified product is made, give information as to the biological role of the product. Other transgenes may be inappropriately expressed as a consequence of the influence of nearby expressed genes. The insertion then acts as a reporter of gene activity, and is a tag for the isolation of those genes. Secondly, transgenes are cell markers, either as hybridization targets, or as producers of cell-autonomous products which can be used to study cell lineage. Finally, the presence of the transgene has been used as a reporter of methylation status of the genome, to examine methylation differences after maternal or paternal transmission and as a possible indicator of genomic imprinting.

5.1 Dominant effects of transgene expression

All expressed transgenes are dominant or co-dominant. In many cases the expression is of no consequence to the host mouse, but in some cases, intentionally or not, the expression has developmental effects.

Considerable work has been done using various oncogenes or proto-oncogenes in transgenics (156,159). Specific tumors can be reproducibly generated by use of particular promoter – enhancer combinations driving appropriate oncogenes. Usually the tumors form in the tissues where the enhancer normally functions.

One dominantly acting transgene which affects early development has been described. Dihydrofolate reductase (DHFR) is an enzyme which converts dihydrofolate to tetrahydrofolate. A mutant DHFR gene, which encodes an enzyme with decreased substrate affinity and turnover, and linked to a strong promoter, has been introduced into a number of mouse lines (160). The enzyme, if over-expressed, will sequester dihydrofolate, so that less is available for synthesis of thymidylate, a DNA precursor. Of the transgenic lines studied, 13 out of 16 had abnormalities which could be ascribed to dominant effects on the mutant DHFR product. Most common were stunted growth and tail abnormalities, but eye and melanocyte defects were also seen. The possibility was raised in Section 4 that these types of defects might be caused by growth retardation. It is probable that in this case they are due to a reduction in growth rate or cell cycle times caused by the defective enzyme. One of the lines is affected in early pre-implantation development. The line is female-sterile. Fertilized eggs from female carriers are retarded in development *in vitro* by 24 h or more, and presumably either do not implant or develop incorrectly post-implantation. Normal eggs fertilized by the transgenic sperm develop into stunted offspring with tail defects. It would seem either that expression of this particular transgene in the oocyte causes a depletion of dihydrofolate in the egg and a retardation of the early cell cycles, or that the transgene is subject to parental imprinting and is expressed in cleavage embryos less when inherited from the sperm than the egg. The former explanation seems more likely since the usual assumption is that paternally inherited imprinted genes are expressed more, rather than less than the maternally inherited (see discussion on imprinting in Section 5.4).

5.1.1 Antisense genes

Constructs are being developed (161) which produce 'antisense' RNA which can hybridize to mRNA and functionally eliminate the message in transgenic animals thereby acting as a dominant mutation. In two reported cases antisense RNA has been demonstrated to reduce the expression of enzymes in pre-implantation mouse embryos. Injecting antisense RNA into the cytoplasm of each blastomere at the four-cell stage reduces the 60-fold increase in activity of β-glucuronidase normally seen at blastocyst by 75% (162). Pronuclear injection of fertilized eggs of a gene, driven by a metallothionin promoter, encoding hypoxanthine phosphoribosyl transferase (HPRT) antisense RNA does not affect the expression of endogenous HPRT, but it will completely abolish expression of a co-injected HPRT minigene (163).

One transgenic line has been reported which produces an antisense transcript which affects development. This is an antisense construct of the myelin basic protein (MBP) gene (164). Absence of MBP mRNA is a characteristic of homozygous *shiverer* mice and produces a trembling phenotype. An antisense MBP gene acts dominantly to produce the trembling phenotype in *shiverer* heterozygotes. The one normal MBP gene in the transgenic animals produces less than half its normal output of mRNA, and the resulting 20% of wild-type MBP mRNA gives the trembling condition.

Over-expressed genes, antisense genes or genes encoding modified products might be used to examine the biology of genes important for early development. However, these experiments may prove problematic as the dominant effect will in many cases kill the embryo and it will not be possible to establish transgenic lines. The injected embryos themselves will have to be examined for abnormalities. There are two ways to avoid the problems this might cause. If the defect is not cell-lethal and does not affect germ cell-production, embryonal stem cells could be used to bring the transgene into the germ line (see Section 6). Another way would be to place the lethal transgene into a mouse line under the control of an element such that it is only transcribed in response to a factor not normally found in mouse cells. Another transgenic line could be established which carries this *trans*-acting factor gene, and expresses it ubiquitously or tissue-specifically. The two lines will be normal transgenic lines, with no ill-effects, but when crossed, 25% of the embryos will inherit both the factor gene and the lethal gene, which will be activated and exert its effect. Several factor/gene systems might be used in this way, including viral and yeast systems (165,166,167).

5.2 Transgenes as probes to identify cell-specific genes

The idea that certain transgenes are affected by their chromosomal location, or that the effect of weak promoters can be strengthened by integration near to certain cellular sequences has given rise to so-called 'enhancer-trap' experiments in transgenic animals. Several experiments have used the *Escherichia coli lacZ* gene which when expressed acts in a cell-autonomous manner, those expressing it stain blue with the substrate 5-bromo-4-chloro-3-indolyl-β-D-galactopyranoside (X-Gal). A *lacZ* gene driven by the P-element promoter has been introduced into *Drosophila* embryos (168). No expression was detected by X-Gal staining in 20% of the embryos. Of the other 80%, about one in four had staining in most or all cells, whilst the rest stained in a restricted manner. Labeling of particular subsets of cells in the nervous system was common, as was staining of parts of the gut. Still others had *lacZ* expression showing a two-segment periodicity, like that of the pair-rule genes. The weak P-element promoter seems to detect cell-specific enhancers or promoters.

The same approach has been applied to mouse embryos. Allen *et al.* (169) injected a *lacZ* gene under control of the herpes simplex virus thymidine kinase (HSV-TK) promoter into fertilized eggs. They examined several hundred embryos by X-Gal staining. Expression was seen in only 11 embryos. In some the blue stain was limited to one tissue, whilst in others it was seen in different, unrelated tissues. One embryo showed staining in somites and forelimb bud, whereas another showed staining in somites, heart, skin and neural tissue. In six of the 11 *lacZ* was expressed in the CNS and three of these had additional expression in other tissues such as heart, somites and lung. Five transgenic lines were established which showed expression in embryonic offspring. One, TkZ710, shows first detectable expression at 11 dpc in a group of cells in the brain, the somites and the neural tube. Later the somite and neural tube expression disappears but staining persists in the brain, in a more restricted region (*Figure 6*).

The assumption of the experiments is that activation of *lacZ* reflects activity of the chromosomal domains into which it has integrated. This remains to be rigorously proved, but the method has potential for identifying genes expressed in particular patterns, candidates for

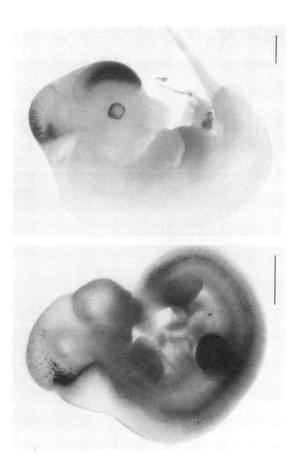

Figure 6.   The photomicrographs show transgenic embryos from the line TKZ710, stained with X-Gal to reveal the expression pattern of *lacZ*. On the left is an 11.5 dpc embryo showing expression in a group of cells in the hind brain and throughout the neural tube and somites. On the right is a 13.5 dpc embryo showing that expression is now seen only in restricted regions of the brain. Scale bars are 1 mm. The photographs were kindly supplied by Nick Allen and are described fully in ref. 169.

regulatory genes or targets of regulatory genes. Furthermore, the genes can be cloned using the *lacZ* gene as a hybridization tag. They may also be mutated by the insertion in the transgenic mice, so that a mutant phenotype could be seen in transgenic mice if it is made homozygous.

The observation that a high proportion of presumably random integration sites show expression in somites and neural tissue should lend some caution to the interpretation of the *in situ* hybridization patterns seen with putative developmental genes such as the *Hox* family.

5.3 Transgenes in the study of lineage

A full understanding of the development of the embryo requires that the developmental origins and relationships between cells be known. Lineages of certain tissues are understood, as described in Section 2, but a great deal remains to be determined, particularly post-implantation. Lineage can be studied in transgenic animals either by using the transgene as a lineage marker, or as a manipulative tool to specifically ablate cells of a particular lineage.

5.3.1 Transgenes as lineage markers

Lineage in the mouse has classically been followed by physically marking single cells or groups of cells, and following them through development. Alternatively chimeras have been made by combining cells of two genetically marked embryos, and looking at the fates of the two cell populations (170). Physically marking single cells presents problems of potential adverse effects of the manipulations and of the presence of the marker on development. Using chimeras presents two problems. First, the chimeras are almost always made pre-implantation, as only then are manipulation and reimplantation procedures practical. Thus lineage information is restricted to pre-implantation fates. Secondly, the combination of genetically distinct strains may not be ideal. If cells of the same genotype have a preference to stay together, or if one genotype has a growth advantage over the other, the contribution of each parental embryo to the resulting tissues will be skewed.

This is not a problem with X-inactivation which produces natural functional mosaicism in all females, and can be used to follow lineage. It has been used, for example to confirm that all cells in a single intestinal crypt are clonal in origin (171). However, the inactivation occurs at random when there are a large number of cells in the embryo and about equal numbers of cells inactivate either X-chromosome. The resulting mosaic is therefore fine-grained, and not very informative as to lineage (170).

The ideal lineage marker would be one which
(i)   marks only one or a few cells;
(ii)   which can mark cells at any stage in embryogenesis;
(iii)   is detectable at the single cell level;
(iv)   confers no growth advantage or disadvantage to the cells it marks.

Retroviral markers fulfil many of these requirements (172).

Pre-implantation mosaics have been generated by infection of 4 – 8-cell embryos with Moloney murine leukemia virus (Mo-MLV). The resulting mice have integrated proviruses, detectable by Southern blotting of DNA from a range of tissues (173). Very few proviral integrants (three out of 52) were common to both the embryo proper and the placenta, which indicates that at the time of integration (possibly 16 cells or later) the embryonic and extraembryonic lineages had largely separated [possibly into inner (embryonic) and outer (extraembryonic) cells]. After birth, each tissue examined separately usually had less than a genome-equivalent of any integrant, therefore the embryo derives from more than one potentially infected cell. Although the level of mosaicism varied from animal to animal it was constant between different organs of a single animal, so extensive cell mixing must occur. The smallest mosaic fraction seen was 12%, or 1/8, indicating that the embryo derives from a pool of no more than eight cells. This can be seen in two ways. One view is that at the time of integration of the proviruses there are eight cells which can go on to form the embryo (perhaps all the eight-cell embryo or the inner cells of the 16-cell embryo). Later, when the epiblast forms the embryo proper, the pool is small enough for descendants of only one ranging to all eight cells to be sampled in different embryos. Alternatively, this epiblast pool may itself be only eight cells.

Interestingly, this work finds that the contribution to germ cells (as defined by looking at transmission of proviral integrants to offspring) is often not the same as to somatic tissues. Several mice were found which had considerable contribution to somatic cells (100% in one case), but had no germ line transmission. In a few cases the explanation might be that the integration inactivates a haploid-required gene (or possibly an imprinted gene expressed early in development), and germ cell contribution is not accurately measured by looking at offspring. Overall, however, the conclusion is that when the germ cell lineage separates from the somatic there are sufficiently few cells present that there is a chance that cells which do not contain viral integrants separate from somatic cells which do. It must also be concluded that the germ line is set aside before the somatic lineages are established to allow the sort of mixing which produces the fine-grained somatic mosaics seen. Animals were also identified of the opposite kind, which have no detectable somatic contribution of certain proviral integrants which are nevertheless in the germ line. These lead to the additional conclusion that the germ cells are set aside very early, when numbers are sufficiently small such that cells with particular integrants can enter the germ line but be excluded from somatic tissue. The idea that germ cells are set apart when cell numbers are small is somewhat controversial (172,174) and must be further substantiated.

Retroviral infection of embryos will prove uniquely valuable when

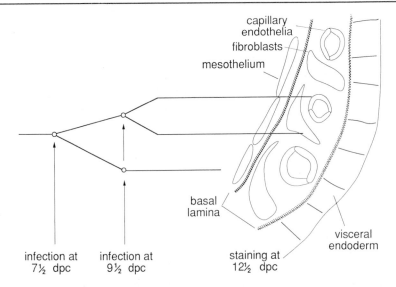

capillary
endothelia

fibroblasts

mesothelium

basal
lamina

visceral
endoderm

infection at   infection at   staining at
7½ dpc   9½ dpc   12½ dpc

Figure 7.   A schematic illustration of the components of the visceral yolk sac and the analysis of mesodermal cell lineages. There are three mesodermally derived cell types; the mesothelium facing the embryo separated by basal lamina from fibroblasts and capillary endothelium cells. Infection of the embryo with a retrovirus carrying the *E.coli lacZ* gene at 7.5 dpc results in clones staining blue with X-Gal at 12.5 dpc which comprise all three cell types, indicating a common progenitor at this time. Infection at 9.5 dpc gives rise to clones of either mesothelium or fibroblasts plus capillaries, demonstrating that the lineages have separated. The data are taken from ref. 175.

applied post-implantation. Identification of single marked cells is possible using retroviruses carrying the *E.coli lacZ* gene under control of a ubiquitously active promoter. Cells containing the integrated provirus will stain blue with X-Gal.

Sanes *et al.* (175) used a replication-defective Mo-MLV containing *lacZ* driven by the SV40 early promoter to infect mouse embryos. They found efficient infection from as early as 7 dpc, which they could subsequently detect by β-galactosidase activity. X-Gal revealed blue-staining cells in clonal groups in the amnion, yolk sac and embryonic skin. After infection at 7 dpc and staining several days later, they found no labeling of the visceral endoderm, but representatives of all mesoderm derivatives in the visceral yolk sac were labeled, indicating a common progenitor at 7 days. Infection 2 days later identified two lineages in the yolk sac mesoderm, only one of which would be labeled in any one clone. One lineage gives rise to fibroblasts and capillary endothelium, the other to mesothelium. Progressive lineage restriction is thus identified (*Figure 7*). They derive the same kind of conclusions for development of the skin.

Although the lineages identified in this experiment were all on accessible surfaces, they also showed that injection of the virus inside the embryonic head allows infection of internal cell lineages.

A potential problem with use of *lacZ* activity as a marker would occur if the gene were not expressed in all cells. If certain cell types did not use the promoter driving *lacZ* these would be falsely excluded from a lineage. Furthermore, expression of the marker gene in the provirus must not be influenced by integration site, which could lead to shutting-down of the gene in certain cells, either randomly or in a lineage-related manner.

5.3.2 Cell-specific ablation

Lineage studies of *C.elegans* have benefited by the use of lasers to ablate specific cells in embryos. A transgenic ablation tool has been described in the mouse, although it differs from laser ablation in certain key ways. The transgenic method uses the diphtheria toxin A-subunit (DT-A) gene linked to a cell-specific promoter (176 – 179). The DT-A protein, if present in cells will, even at extremely low concentrations, kill the cell, but in the absence of the B-subunit cannot enter any other cell. Tissue-specific transcription of the DT-A gene leads to tissue-specific cell death mediated by the toxin. In theory any cell in which that promoter is active will be ablated, and the subsequent effects on development seen. If the ablated cells are not essential, transgenic lines can be established with the ablated phenotype.

Several lines of mice with microphthalmia have been produced using the DT-A gene driven by the γ-crysallin promoter, which is active specifically in the lens (178). Similarly, dwarf mice are produced by a construct in which DT-A is fused to the growth hormone promoter (179). The animals are much smaller than normal, have no detectable circulating growth hormone (GH), and reduced pituitaries. The pituitaries do have a few cells containing GH detectable by immunohistochemistry but these represent only about 0.01% of the normal number. Lineage of GH-producing and prolactin (PL)-producing cells has been deduced in these experiments. It was thought that both cell types have a common progenitor, which produces both hormones. If this were so there should also be no PL-positive cells in the pituitary of the dwarf mice. In fact there are such cells, although their numbers have been much reduced (but not as much as the GH cells). The simplest conclusion is that there is a common progenitor for most PL-producing cells which also expresses GH, but some PL cells must derive from a progenitor in which the GH gene is inactive. An alternative is that the two cell types do not share a progenitor, but the development of PL-positive cells is dependent in some way on the GH cells.

Although they have impaired fertility, the dwarf mice can be maintained as a line. A third ablation experiment reported (177) results in neonatal death of the transgenic animals. This uses the rat elastase-I promoter driving DT-A, which is specific for the acinar cells of the pancreas. Embryos into which the gene has integrated lack, wholly or partially, a pancreas. Whilst the absence of the elastase-expressing acinar cells was expected, the lack of duct and islet cells suggests either that these cells will not develop in the absence of acinar cells, or that precursors of the two cell types express, transiently, the elastase-I gene (and hence the DT-A protein).

The inability to transmit the construct means that all these observations were done on the injected embryos, and the basis of the variability seen between different embryos cannot be traced. There are several ways which would allow an embryonic lethal transgene to be passed on. One would be to use a repressed DT-A gene, only activatable when crossed with another strain containing a *trans*-acting factor. Another is to use mosaic animals. These might be found among the initially injected transgenics and will have the transgene in only a proportion of the cells, excluded from those tissues where it is expressed, but present in the germ line from where it can be passed on. Mosaics (or chimeras) could also be made by introducing DT-A containing embryonal stem cells into normal blastocysts.

Such mosaics or chimeras will allow the two alternatives raised by the dwarf mice and the apancreatic mice to be distinguished. If the development of duct and islet cells in the pancreas requires the presence of acinar cells, in mosaics the acinar cells will develop from normal cells, and should, on this model, allow transgenic duct and islet cells to develop even though they contain the elastase/DT-A construct. Absence of transgenic duct and islet cells would suggest that they have an elastase-expressing progenitor. Identification of transgenic cells within a population might be possible using *in situ* hybridization to the transgene, but better would be integration of a *lacZ* into the transgenic cells and staining with X-Gal.

The crucial difference between this type of ablation experiment and the *C.elegans* laser ablation is that it does not allow other cells to join a lineage; any which do will also be ablated. It is, therefore, far less precise a tool (176) for lineage analysis. It is nevertheless a useful additional approach for examining the development of certain tissues in the absence of others, and may be fruitfully applied to early development in the future.

5.4 Genomic imprinting and transgenes

An otherwise identical chromosome complement behaves differently if inherited from the father rather than the mother. It is possible to replace either pronucleus of the fertilized egg with pronuclei from other embryos. In this way it has been shown that embryos with two female pronuclei (gynogenetic embryos) or with two male pronuclei (androgenetic embryos) are not viable (180–182). The two types of manipulated embryo, furthermore, die with different abnormalities. Gynogenetic embryos fail to develop adequate extraembryonic tissues, although the embryo itself is morphologically normal. Androgenetic embryos have apparently normal extraembryonic structures, but have much reduced embryonic material

(180). Not only are both parental genomes required for normal development, but the absence of one or the other genomes results in reciprocal defects.

The requirement that nuclei derive from both parents does not extend over the whole genome, but can be localized to particular chromosomal regions. By exploiting the tendency of heterozygotes carrying Robertsonian chromosome translocations to have meiotic non-disjunctions, and so sometimes to produce gametes with two or no copies of a particular chromosome, Cattanach, Searle, Lyon and colleagues have produced embryos with a normal chromosome complement, but one of which is inherited solely from one parent [known as maternal or paternal disomics (*Figure 8*)]. They have shown that for some chromosomes, for example 1, 4, 5, 9, 13, 14, and 15, uniparental disomics develop normally (182,183), whilst for others, such as 7, 8 and 17, they do not survive to term. In some cases the lethalities are unidirectional. A paternal disomic for chromosome 6, for example, is normal, but a maternal disomic dies before birth (184). Other uniparental disomics survive to term but with visible phenotypes. Neonates paternally disomic for chromosome 11 are larger than normal, whilst those with both chromosome 11s of maternal origin are smaller than normal littermates (183). Intrachromosomal translocations have been employed to show an imprinting effect localized to the distal part of chromosome 2. Paternal disomics are short and squat and are hyperactive. The maternal disomic neonates have long bodies, arched backs and are almost totally inactive (183). Both die soon after birth. The reciprocal nature of the phenotypes of both examples above is striking.

Some mutations show evidence of imprinting. *Hairpin tail* (T^{hp}) is a small deletion in the *t*-complex region. It produces a dominant, viable effect on tail morphology if inherited from the male, but a much more severe defect leading to prenatal death if the gene is transmitted from the mother (185). One of the *t*-chromosomes, t^{wLub2}, has a smaller deletion, within the T^{hp} deletion, which shows the same maternal effect, and narrows down the imprinted segment to perhaps a single locus (186). McGrath and Solter (187) showed by nuclear transfer that the maternal lethality is a nuclear defect rather than a defect in egg cytoplasm.

The presumption is that certain parts of the genome are genetically imprinted by passage through egg or sperm and this affects the subsequent activity in embryogenesis of some genes. The reciprocity of the phenotypes suggests that the paternal genome is more active in the embryo proper (although it may be the opposite in extraembryonic tissue). It is difficult to know how many genes are involved, as most experiments so far have used whole chromosomes or large parts of it. Some have suggested (188) that 25% of the genome is imprinted, but there is no reason to believe as yet that there is more than one gene per chromosomal region which produces an imprinted phenotype [which stands at eight at

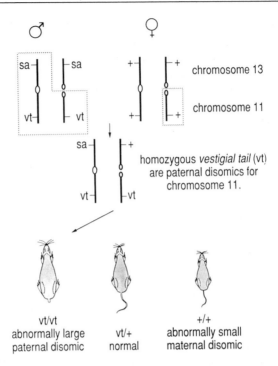

Figure 8.   Generation of chromosome 11 uniparental disomic mice. Animals heterozygous for a Robertsonian translocation which fuses chromosomes 11 and 13 frequently produce gametes with two or no copies of either chromosome. An embryo resulting from a gamete with two and a gamete with none will have the correct chromosomal complement, except that one chromosome pair will be derived from only one parent. Shown here is the derivation of an animal with paternal disomy for chromosome 11. The morphological marker *vestigial tail* (vt) permits identification of such animals as homozyotes. The mutation *satin* (sa) is a marker for chromosome 13. The cross performed as shown results in ~1–2% of animals paternally disomic for chromosome 11 which have vestigial tails and are larger than their normal littermates. Animals homozygous for the wild-type allele are maternal disomics and are abnormally small. The reverse cross, with females carrying the Robertsonian chromosome with *vestigial tail*, produces abnormally small maternal disomics which are vestigial tail. Data are taken from ref. 183.

present 1891)]. On the other hand the genetic test for imprinting only detects those genes which have a lethal or visible phenotype. Many genes may have different activities during early development, depending on parental source, but may be developmentally silent. Furthermore, the whole genome may bear the mark of imprinting, but not all genes may respond to the imprint by a difference in expression.

It is not known what, if any, the function of imprinting might be. One possibility is that it has evolved as a way of preventing parthenogenesis, of ensuring that a sexual mode of reproduction is maintained.

Alternatively, the phenomenon might be largely a by-product of the way DNA is treated during spermatogenesis or oogenesis, which is only detected when the embryos are experimentally manipulated (and which might nevertheless serve to prevent parthenogenesis).

A physical basis for imprinting is sought. The imprinted genome must, presumably, maintain its state at least until the embryo proper begins to develop, through many rounds of chromosome replication. The imprint could conceivably be chromosomal proteins which are added to the genome during gametogenesis, and which are passed on to daughter chromosomes at replication, but do not transfer to the homologous chromosome inherited from the other parent. Perhaps more likely would be a modification of the DNA itself, and recently the methylation state of the genome has been examined in this light.

It is the cytosine residue (C) which is methylated in mammalian DNA, and methyl-C is found almost always in the dinucleotide CG. The restriction enzymes *Hpa*II and *Msp*I cleave the sequence CCGG, but only *Msp*I will cut the site if the second C is methylated. The enzymes together will identify methyl-Cs occurring in genomic DNA at CCGG sites. A transgene provides a molecular probe which can be inherited in a hemizygous manner from either parent, and the restriction enzymes can be used to examine the methylation status of CCGG sequences within and flanking the transgene by probing Southern blots. Reik *et al.* (190) showed that in one case of seven examined, a transgene comprising a chloramphenicol acetyl transferase (CAT) gene with an immunoglobulin (Ig) enhancer, was methylated at several CCGG sites within the transgene in mid-gestation embryos if it was inherited from the mother, but was much less methylated following paternal inheritance. The methylation pattern was reversible. If the transgene was passed through the egg or sperm it would gain or lose methylation. The methylation of the transgene persisted through to the adult in all tissues except the testes, where a maternally inherited (methylated) transgene loses its methylation during spermatogenesis. The transgene was not expressed anywhere.

A number of other transgenics were examined by Sapienza *et al.* (191) who looked at the methylation status of troponin-I transgenes in the tails of adult mice. Of five examined, one showed no methylation difference, three had more methylation when inherited from the mother, and one was the reverse, being less methylated after maternal inheritance. The methylation in these cases did not affect their expression, all were expressed as expected in muscle irrespective of parental origin. By contrast, another transgene has been shown to have expression differences depending on origin (192). It is made up of elements from Rous sarcoma virus, the *Ig*α gene and the *c-myc* oncogene. The transgene is less methylated at CCGG sites when inherited from the father, and is expressed in the adult myocardium. The maternally transmitted transgene is not expressed, and is more methylated. This transgene and the others above are capable of switching methylation status when sex of parental source is changed.

One transgene has been described which will only switch one way (193). On passing down the male line this transgene (derived from the Hepatitis B genome) is undermethylated and is expressed in the liver. However, when it is passed through the female it becomes methylated and no longer expresses. The methylation state is now irreversible, so that male offspring from a transgenic mother pass on a methylated, unexpressed, transgene.

It is assumed that the transgene is detecting methylation events which happen normally during gametogenesis. However, the identification of a phenomenon which has the hallmarks of imprinting does not prove that it is related. The methylation data do have some striking features. The imprinted transgenes are usually, but not always, undermethylated when passed through the sperm. This contrasts with the higher total methylation of sperm versus oocyte DNA (194,195). Expressed genes are normally undermethylated, and so undermethylation of paternally inherited transgenes is consistent with the hypothesis that it is the paternally imprinted genome which is more active. However, complete lack of expression of a gene which is maternally inherited probably does not occur naturally and the irreversible methylation and inactivation of the Hepatitis B transgene is something which must be abnormal. Placing a transgene within a chromosome is a disruptive event, and the effect this has on chromatin structure and behavior is not known. It is also not known if these methylation differences occur normally along chromosomes. It should certainly be possible to do similar experiments to look at methylation of natural restriction fragment length variants passed between inbred mice.

It must again be emphasized that the physical imprinting seen as methylation is not necessarily related to the genetically detected imprinting. It if is related, then it is not necessarily the primary imprint, but may be a secondary event which reveals the imprint. It is possible, for example, that the primary imprint is made by binding of proteins, which consequently affect the action of methylases on the DNA. In any case, identifying the nature of imprinting will at the very least tell us a lot about gametogenesis, and possibly about gene expression in early development.

6. Embryonal stem cells: manipulating the genome

Paucity of material from early embryos led to the development of model systems, which were initially used for the study of early embryogenesis but have more recently found application as a means of manipulating the germ line of mice. These cell systems are embryonal carcinoma (EC) or teratocarcinoma cells and EK or embryonal stem (ES) cells.

6.1 Embryonal carcinoma cells

Teratomas and teratocarcinomas (the former are benign, the latter malignant tumors) are found both in humans (196) and mice (197). These tumors are a disorganized mass of cells within which can be identified a number of different, differentiated types, such as cartilage, skin, neural tissue, bone and muscle. Teratocarcinomas occur spontaneously in some mouse strains (197), but can be made from any strain by implanting embryonic material from 7.5 dpc or earlier under the kidney capsule of syngeneic hosts (198). A tumor develops which can be transplanted or explanted into culture. Once in culture certain cells from the tumor proliferate. These are stem cells, and resemble cells of the blastocyst. If treated appropriately in culture most EC cells will differentiate into the wide range of cell types seen in tumors (197). Some lose the capacity to differentiate, whilst others will only give rise to a restricted range of cell types. For example, under the right conditions F9 cells will form visceral or parietal endoderm but no other differentiated products (35).

If EC cells are microinjected into the blastocoel cavity of the blastocyst they come under the influence of developmental signals and usually participate in normal development (197,199). The resulting chimeric offspring have tissues composed of both host embryo and donor EC cells. In one reported case (199) EC cells also contributed to the germ line of a chimera, which passed the EC genome on to its progeny.

EC cells have proven invaluable for studying molecular and cellular aspects of early development. The recent work on ES cells, however, has greatly increased the possibilities and I will concentrate on this system for the rest of the section.

6.2 Embryonal stem cells: growth *in vitro*

By a number of criteria, tumor-derived EC cells appear equivalent to epiblast cells of the early post-implantation blastocyst (200). Methods have been described (200 – 202) by which blastocysts can be explanted directly into culture and stem cells derived from the ICM. The blastocyst may be explanted after normal development or after a period of delayed implantation and growth *in utero*. Some initial experiments physically separated the components of the blastocyst and seeded the ICMs into culture (201), but ICMs will proliferate on trophoblast cells attached to a solid support and can be removed later for culture.

Most methods use a feeder layer of fibroblasts to maintain the undifferentiated stem cell characteristics. Feeders have been substituted effectively by medium conditioned by buffalo rat liver (BRL) cells (203).

ES cells in culture usually keep a normal karyotype, with the exception of the X-chromosome in female lines. The time at which ES cells are isolated precedes the time of normal X-inactivation. Consequently both X-chromosomes are probably active in these cells, and this is possibly detrimental. The cells lose all or part of one X-chromosome (204).

Translocations and loss of a Y-chromosome from male lines have also been observed (205).

Stem cells can be induced to differentiate *en masse* simply by growth in suspension (202). The cells aggregate and undergo differentiation so that a layer of endoderm forms around a core. These are simple embryoid bodies; models for the endoderm and ectoderm of the early post-implantation blastocyst. Continued culture in suspension allows further differentiation into cystic embryoid bodies which swell up with accumulated fluid and differentiate an ectodermal layer.

If simple embryoid bodies are replated onto a solid substrate they attach and outgrow, forming terminally differentiated products such as cartilage, epithelia and neuronal tissue (202).

6.2.1 ES cells derived from developmental lethal mutants

ES cells can be used as a tool for study of homozygous lethal mouse mutants, providing that the mutants develop at least as far as blastocyst.

Magnuson *et al.* (206) derived ES lines from a mating of mice carrying the *t*-chromosome, t^{w5} (see Section 4.3). Homozygous t^{w5} embryos are abnormal from the egg cylinder stage, but one of three ES lines derived was homozygous as judged by chromosomal, protein and DNA markers. The cells were viable in culture and could give rise in culture and in tumors to a diverse set of differentiated cell types. The conclusion is that t^{w5} is not a simple cell-lethal, but must block a specific developmental event.

By contrast, Niswander *et al.* (124) were unable to derive ES lines from two different homozygous *albino* deletions (see Section 4.2). Lines established from heterozygous mating were genotyped with DNA probes, and no line was found homozygous for deletions as far as the extraembryonic ectoderm function, nor for smaller deletions encompassing the embryonic ectoderm function (*see Figure 4*). The deviation from the expected number of lines was highly significant. A gene defined by the smaller of the deletions must be essential for survival of ES cells.

6.3 Development of ES cells in chimeras

If ES cells are injected into a syngeneic adult host, they will form a teratocarcinoma. The real power of ES cells, however, lies in their behavior in chimeras. If the cells are injected into blastocysts (or possibly aggregated with a morula stage embryo) the cells participate fully in normal development (207). Pigmented ES cells in an *albino* host give chimeras easily identified by their mixture of pigmented and unpigmented hair. Chimerism in the dermis can be seen if host and donor differ at the *agouti* locus, which acts in the dermis to affect adjacent melanocyte pigmentation (208). Chimerism in other tissues is detectable by isozyme analysis or Southern blotting of DNA.

ES cells generally form extensive chimeras, with as much as 75% or more donor-derived tissue. Chimerism extends to the germ line (207).

Female cells (XX or XO) in female hosts form chimeras which have a proportion of ES-derived oocytes. More productive is the use of male ES cells. As testis-determination is dominant, male cells injected into a female embryo will, if their contribution is high enough, push the embryos into male development. As sperm can only form from XY cells these chimeras will make sperm entirely from ES-derived cells. Chimeras made using XY ES-cells have a distorted sex ratio, some females having been converted to males. Some males have a chimeric germ line (those derived from male host blastocysts) whilst others (derived from female hosts) pass on only the ES genome.

The contribution of ES lines made from developmentally mutant embryos to chimeras will be informative. Exclusion of the mutant cell lines from certain tissues will indicate a cell-autonomous defect. The contribution to each tissue can be checked on a gross level by isozyme or DNA analysis. A finer analysis requires a cell-autonomous marker, which could be a *lacZ* gene, introduced into the cells in culture, and identified in the chimeras by X-Gal staining.

6.4 Manipulation of the genome

The availability of cultured cells whose genomes can be altered and passed through the germ line into mutant mice is one of the most important recent advances of developmental biology. The great advantage of ES cells is that large numbers can be manipulated and tested before cloning and injection into blastocysts. *Figure 9* outlines various steps which can be taken to manipulate the genomes of ES cells.

6.4.1 Transgenic ES cells

Transgenic animals can be made after introduction of DNA into ES cells. The DNA is introduced into the cells by standard techniques: calcium phosphate, electroporation or microinjection (209). In one such experiment, a human $\alpha 1$(II)-collagen gene, linked to a neomycin resistance (neo^R) gene has been put into ES cells by calcium phosphate precipitation (210). Growth in G418 selected those cells which had taken up the DNA. Two clones which had $5-10$ copies of the human gene were identified. Only very low transcription of the gene was seen in stem cells (less than one molecule per cell). When introduced into chimeras the human collagen gene was active in only those tissues which normally express mouse type II collagen, and the human protein could be detected in the tissues by antibody staining.

Transgenic ES cells can be characterized before making chimeras. If the transgenes are tightly regulated a gene expected to have a lethal effect on a particular cell type or effect aberrant cellular interaction (as discussed in Section 5) could be introduced into chimeric animals in this way. Particular lethalities might exclude the cells from certain tissues, in which case wild-type host cells will compensate. As long as the lethality is not

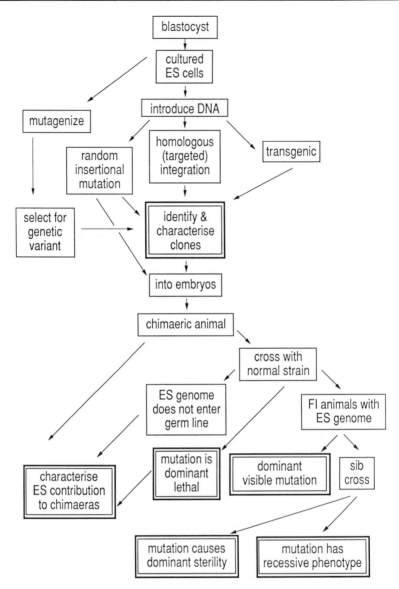

Figure 9. Potential manipulations and analyses of embryonal stem cells. Double boxes indicate key analyses. Characterization of manipulated ES clones *in vitro* is an important feature, which distinguishes use of ES cells from microinjected embryos to manipulate the germ line.

expressed in germ cells, the transgene will be passed on through the germ line where any dominant lethality can be studied in the next generation. The same modified cell line can be used many times, so that the experiments are reproducible and controlled in a way that DNA injection into embryos does not permit.

6.5 Mutagenesis in ES cells

ES cells can be mutated in the same way as other cultured cells (211). Insertional mutations can be made using retroviruses, for example. This has been carried out using replication-defective viruses carrying a neo^R gene (212). The mutated cells have been used to look for insertional mutations in the mice derived from them. The efficiency of mutagenesis is probably the same as with retroviral infection of embryos, but the rate at which inserts can be tested for phenotype is much increased. Multiple retroviral insertions have been obtained, with as many as 20 integrants per cell. If unselected cells from culture are injected into blastocysts, several cells contribute to the germ line and mice transmitting up to 40 different proviruses have been seen (212). It is, therefore, an efficient way of searching for newly induced dominant visible or recessive mutations. Sib-crossing the F1 generation and looking for each provirus by Southern blotting DNA from the F2 will quickly identify those never seen as homozygotes (and therefore causing recessive-lethals). Mutations of this type, and also visible mutations, have been associated with particular proviral insertions (E.Robertson, personal communication).

One possible way of increasing the efficiency of mutagenesis is to use a promoter or enhancer 'trap'; that is to introduce into the cell a dominant selectable marker with no promoter or a weak promoter (see Section 5.2). Cell clones growing after selection should have the marker lying in (and causing a mutation in) an active gene. This type of experiment has been done both in EC cells (213) and in ES cells (D.McLeod, R.Lovell-Badge and I.J.Jackson, unpublished data). Cloned DNA flanking the inserts hybridizes to transcripts in stem cells. These insertions are expected to cause a particular class of mutation; early pre- or post-implantation recessive lethals, but dominant, haplo-insufficient, effects and other recessive phenotypes might be seen.

Selection of ES cells for biochemical variants is possible. HPRT-deficient cells are unable to metabolize 6-thioguanine, and are resistant to its toxicity. The HPRT gene is on the X-chromosome, so male cells are readily mutated. Kuehn et al. (205) and Hooper et al. (214) independently isolated HPRT-deficient ES lines, the first from retrovirally mutated cell populations, the second selected a spontaneous mutation in cells grown feeder-free on BRL-conditioned medium. Both groups have introduced the cells into germ line chimeras, and derived carrier females and homozygous mutant males. The males are completely deficient for HPRT activity. Although the deficiency is a model for the serious human neurological disorder, Lesch – Nyhan syndrome, the mice are completely viable and healthy. Nevertheless, this represents an important advance, in that mice have been produced of a pre-determined genetic composition by direct manipulation of the genome of EC cells.

Other selection procedures might be devised, using other biochemical, morphological or immunological criteria (211). However, convenient selection schemes are not possible for most genes of interest. In addition most genes are autosomal, and the treatment required to give two hits might be so detrimental to the ES cells as to render them incapable of forming germ line chimeras.

6.6 Targeted mutagenesis

An alternative to selection for function is to specifically target a mutation to an autosomal gene by homologous recombination, and to identify ES cell clones which are heterozygous for an insertion into the gene. When contributing to the germ line of chimeras they will transmit a wild-type copy to half the derivative sperm and the mutant to the other half.

In yeast, most integration of linear DNA occurs by recombination into the homologous chromosomal gene (215). Methods which use linear DNA to promote homologous recombination into mammalian cells have been developed. The efficiency of unselected, homologous recombination is of the order of 1 in 100 to 1 in 10 000 integration events, so methods need to be used to aid identification of the targeted clones.

Smithies *et al.* (216) have used electroporation to introduce DNA into several types of human cells. The DNA molecules they have used contain β-globin linear ends with a neo^R gene and an *E.coli.* suppressor tRNA gene. These could serve as markers to demonstrate homologous recombination into the chromosomal β-globin gene. Cells which had taken up DNA were selected by growth on G418, and all the integration sites cloned using selection for the suppressor function in *E.coli*. A proportion of the plaques hybridized both to probes within the targeting construct and probes flanking the predicted integration site. About 1 in 1000 of the plaques contained the flanking DNA from the β-globin locus. It proved possible to clone cells containing the homologously recombined gene by sib-selection from pools.

Model systems have also been used to test conditions for recombination by Cappechi and colleagues (21). These workers generated homologous recombinants between two defective neo^R genes (217). They also have used the technique to target foreign DNA into and mutate the HPRT gene in ES cells (218), using a construct containing a neo^R gene and linear ends from the HPRT locus. All cells taking up DNA are first selected by growth on G418, followed by growth on 6-thioguanine to select the HPRT-deficient cells. They calculate the efficiency of targeting as ranging from 1 in 40 000 cells taking up DNA, to 1 in 1000, depending on the length of homology between the targeting construct and the chromosomal gene.

A much higher efficiency has been reported (219) for the repair of a defective HPRT gene in ES cells. In this case no neo^R gene was used, instead selection was carried out on HAT-medium for restored HPRT function. An accurate estimate of efficiency is not possible, but by comparison with the efficiency by which a neo^R gene alone gave growth on G418 medium, they suggest an upper limit of 14%, although this is likely to be an over-estimate.

It will be necessary to increase the efficiency in order to identify integrations into genes for which selection techniques are not available. Alternatively, a way must be devised to screen many clones quickly and easily. If the targeted gene is active in ES cells, promoterless neo^R genes can be used to eliminate the majority of insertions in inactive genes, and reduce the background by at least 10-fold (218, I.J.Jackson and R. Lovell-Badge, unpublished results). Decreasing the length of DNA in the construct which is not homologous to the chromosomal integration site may also increase efficiency.

Rapid screening of clones is facilitated by the PCR (14). The method uses a pair of opposing oligonucleotides to prime multiple, successive rounds of DNA replication and to amplify the DNA between them by many orders of magnitude. A primer within the targeting construct (but absent from the locus), in combination with a nearby primer which is not in the construct, will only amplify DNA if integration occurs into the locus. PCR can be carried out on minute quantities of DNA, which does not have to be purified from cell lysates, and takes only a few hours. Many hundreds of clones can be screened, either individually or in pools.

After a targeted ES clone is isolated, the cells have to be injected into blastocysts to make germ line chimeras in which the mutated gene is passed on to the next generation. Half the ES-derived offspring will carry the mutation as heterozygotes and so sib-crosses will generate homozygotes, which can be tested for effects of the mutation. The method should enable mutations to be made in any gene of intererst. Particular targets currently under examination are the *Hox* genes, other *Drosophila* homologs, and proto-oncogenes.

In conclusion, the approaches using ES cells are very powerful. Above all this technology holds out the most exciting prospects for the future of molecular and developmental genetics in the mouse.

7. Acknowledgements

I am a Fellow of the Lister Institute for Preventive Medicine. I thank the Lister Institute, the MRC and the Royal Society for research and travel support. I also thank many colleagues for discussions and communication of data prior to publication, in particular Nick Hastie, Robin Lovell-Badge, Liz Robertson, Duncan Davidson and Bob Hill. Stephen Gaunt and Nick Allen provided *Figure 3a* and *b*, for which I am grateful. Excellent photographic and graphic assistance was provided by Norman Davidson, Douglas Stuart and Sandy Bruce, and library services by Sheila Mould and Helen Moffat.

8. References

1. Hogan,B.H, Costantini,F. and Lacy,E. (1986) *Manipulating the Mouse Embryo, A Laboratory Manual.* Cold Spring Harbor Laboratory Press, Cold Spring Harbor, New York.
2. Pratt,H.P.M. (1987) Isolation, culture and manipulation of preimplantation mouse embryos. In *Mammalian Development: A Practical Approach.* Monk,M. (ed.), IRL Press, Oxford, pp. 13–42.
3. Willasden,S.M. (1986) Nuclear transplantation in sheep embryos. *Nature,* **320,** 63–65.
4. Kelly,S.J. (1977) Studies on the developmental potential of 4- and 8-cell stage mouse embryos. *J. Exp. Zool.,* **200,** 365–376.
5. McLaren,A. (1976) *Mammalian Chimaeras.* Cambridge University Press, Cambridge.
6. Johnson,M.H. and Ziomek,C.A. (1981) The formation of two distinct cell lineages within the mouse morula. *Cell,* **24,** 71–80.
7. Keynes,R.J. and Stern,C.D. (1988) Mechanisms of vertebrate segmentation. *Development,* **103,** 413–429.
8. Hogan,B.H., Holland,P.W.H. and Schofield,P.N. (1985) How is the mouse segmented? *Trends Genet.,* **1,** 67–74.
9. Tam,P.P.L. and Meier,S. (1982) The establishment of a somitomeric pattern in the mesoderm of the gastrulating mouse embryo. *Am. J. Anat.,* **164,** 209–225.
10. Tam,P.P.L., Meier,S. and Jacobson,A.G. (1982) Differentiation of the metameric pattern in the embryonic axis of the mouse. III. Somitomeric organization of the presomitic mesoderm. *Differentiation,* **21,** 109–122.
11. Monk,M. (1987) Biochemical microassays for X-chromosome-linked enzymes HPRT and PGK. In *Mammalian Development: A Practical Approach.* Monk,M. (ed.), IRL Press, Oxford, pp. 139–169.
12. Howlett,S.K. (1987) Qualitative analysis of protein changes in early mouse development. In *Mammalian Development: A Practical Approach.* Monk,M. (ed.), IRL Press, Oxford, pp. 163–181.
13. Piko,L. and Clegg,K.B. (1982) Quantitative changes in total RNA, total poly(A) and ribosomes in early mouse embryos. *Dev. Biol.,* **89,** 362–378.
14. Saiki,R.K., Gelfand,D.H., Stoffel,S., Scharf,S.J., Higuchi,R., Horn,G.T., Mullis,K.B. and Erlich,H.A. (1988) Primer-directed enzymatic amplification of DNA with a thermostable DNA polymerase. *Science,* **239,** 487–491.
15. Chelly,J., Kaplan,J.-C., Maire,P., Gautron,S. and Kahn,A. (1988) Transcription of the dystrophin gene in human muscle and non-muscle tissues. *Nature,* **333,** 858–860.
16. Rappolee,D.A., Brenner,C.A., Schultz,R., Mark,D. and Werb,Z. (1988) Developmental expression of PDGF, TGF-α and TGF-β genes in preimplantation mouse embryos. *Science,* **241,** 1823–1825.
17. Watson,C.J. and McConnell,J. (1987) Construction of cDNA libraries for preimplantation mouse embryos. In *Mammalian Development: A Practical Approach.* Monk,M. (ed.), IRL Press, Oxford, pp. 183–197.
18. Taylor,K.D. and Piko,L. (1987) Patterns of mRNA prevalence and expression of B1 and B2 transcripts in early mouse embryos. *Development,* **101,** 877–892.
19. Pratt,H.P.M., Bolton,V.N. and Gudgeon,K.A. (1983) The legacy from the oocyte and its role in controlling early development of the mouse embryo. *CIBA Found. Symp.,* **98,** 197–227.
20. Flach,G., Johnson,M.H., Braude,P.R., Taylor,R.A.S. and Bolton,V.N. (1982) The transition from maternal to embryonic control in the two-cell mouse embryo. *EMBO J.,* **1,** 681–686.
21. Braude,P.R., Bolton,V.N. and Moore,S. (1988) Human gene expression first occurs between four and eight cell stages of preimplantation development. *Nature,* **332,** 459–461.
22. Crosby,I.M., Gandolfi,F. and Moor,R.M. (1988) Control of protein synthesis during early cleavage of sheep embryos. *J. Reprod. Fert.,* **82,** 769–775.
23. Howlett,S.K. (1986) A set of proteins showing cell-cycle dependent modification in the early mouse embryo. *Cell,* **45,** 387–396.
24. Bensaude,O., Babinet,C., Morange,M. and Jacob,F. (1983) Heat-shock proteins, the first major products of zygotic gene activity in the mouse embryo. *Nature,* **305,** 331–333.

25. Houliston,E.S., Guilly,M.-N., Courvalin,J.-C. and Maro,B. (1988) Expression of nuclear lamins during mouse preimplantation development. *Development, **102***, 271–278.

26. Bayna,E.M., Shaper,J.H. and Shur,B.D. (1988) Temporally specific involvement of cell-surface β-1,4,galactosyltransferase during mouse embryo morula compaction. *Cell*, **53**, 145–157.

27. Watson,A.J. and Kidder,G.M. (1988) Immunofluorescence assessment of the timing of appearance and distribution of Na/K-ATPase during mouse embryogenesis. *Dev. Biol.*, **126**, 80–90.

28. Sobel,J.S. and Allegro,M.A. (1985) Changes in the distribution of a spectrin-like protein during development of the preimplantation mouse embryo. *J. Cell Biol.*, **100**, 333–336.

29. Schatten,H., Cheney,R., Balczon,R., Willard,M., Cline,C., Simerly,C. and Schatton,G. (1986) Localisation of fodrin during fertilization and early development of sea urchins and mice. *Dev. Biol.*, **118**, 457–466.

30. Wu,T.-C., Wan,Y.-J., Chung,A.E. and Damjanov,I. (1983) Immunohistochemical localisation of entactin and laminin in mouse embryos and fetuses. *Dev. Biol.*, **100**, 496–505.

31. Peyrieras,N., Hyafil,F., Louvard,D., Ploegh,H.L. and Jacob,F. (1983) Uvomorulin; a non-integral membrane protein of early mouse embryos. *Proc. Natl. Acad. Sci. USA*, **80**, 6274–6277.

32. Shirayoshi,Y., Okada,T.S. and Takeiki,M. (1983) The calcium dependent cell–cell adhesion system regulates inner cell mass formation and cell surface polarisation in early mouse development. *Cell*, **35**, 631–638.

33. Lee,S., Gilula,N.B. and Warner,A.E. (1987) Gap junctional communication and compaction during preimplantation stages of mouse development. *Cell*, **51**, 851–860.

34. Wilkinson,D.G., Bailes,J.A., Champion,J.E. and McMahon,A.P. (1987) A molecular analysis of mouse development from 8 to 10 days *post coitum* detects changes only in embryonic globin expression. *Development*, **99**, 493–500.

35. Hogan,B.L.M., Barlow,D.P. and Tilly,R. (1983) F9 teratocarcinoma cells as a model for the differentiation of parietal and visceral endoderm in the mouse embryo. *Cancer Surv.*, **2**, 115–140.

36. Meehan,R.R., Barlow,D.P., Hill,R.E., Hogan,B.L.M. and Hastie,N.D. (1984) Pattern of serum protein gene expression in mouse visceral yolk sac and fetal liver. *EMBO J.*, **3**, 1881–1885.

37. Chase,D.G. and Piko,L. (1973) Expression of A and C-type particles in early mouse embryos. *J. Natl. Cancer Inst.*, **51**, 1971–1973.

38. Ono,M., Cole,M.D., White,A.T. and Huang,R.C. (1980) Sequence organisation of cloned intracisternal A particle genes. *Cell*, **21**, 465–473.

39. Piko,L., Hammens,M.D. and Taylor,K.D. (1984) Amounts, synthesis and some properties of intracisternal A-particle related RNA in early mouse embryos. *Proc. Natl. Acad. Sci. USA*, **81**, 488–492.

40. Vasseur,M., Conamine,H. and Duprey,P. (1985) RNAs containing B2 repeat sequences are transcribed in the early stages of mouse embryogenesis. *EMBO J.*, **4**, 1749–1753.

41. Adamson,E.D. (1987) Oncogenes in development. *Development*, **99**, 449–471.

42. Wilkinson,D.G., Bailes,J.A. and McMahon,A.P. (1987) Expression of the protooncogene *int-1* is restricted to specific neural cells in the developing mouse embryo. *Cell*, **50**, 79–88.

43. Wilkinson,D.G., Peters,G., Dickson,C. and McMahon,A.P. (1988) Expression of the FGF-related protooncogene *int-2* during gastrulation and neurulation in the mouse. *EMBO J*, **7**, 691–695.

44. Propst,F., Rosenberg,M.P., Iyer,A., Kaul,K. and Vande Woude,G.F. (1987) *c-mos* protooncogene RNA transcripts in mouse tissues; structural features, developmental regulation and localisation in specific cell types. *Mol. Cell. Biol.*, **7**, 1629–1637.

45. Muller,R., Slamon,D.J., Tremblay,J.M., Cline,M.J. and Verma,I.M. (1982) Differential expression of cellular oncogenes during pre- and post-natal development of the mouse. *Nature*, **299**, 640–644.

46. Slamon,D.J. and Cline,M.J. (1984) Expression of cellular oncogenes during embryonic and fetal development of the mouse. *Proc. Natl. Acad. Sci. USA*, **81**, 7141–7145.

47. Zimmerman,K.A., Yancopoulos,G.D., Collum,R.G., Smith,R.K., Kohl,N.E., Denis, K.A., Nau,M.M., Witte,O.N., Toan-Allerand,D., Gee,C.E., Minna,J.D. and Alt,F.W. (1986) Differential expression of *myc* family genes during murine development. *Nature*, **319**, 780–793.

48. Jakobovits,A., Schwab,M., Bishop,J.M. and Martin,G.R. (1985) Expression of *N-myc* in teratocarcinoma stem cells and mouse embryos. *Nature,* **318,** 188–191.

49. Villares,R. and Cabrera,C.V. (1987) The *achaete-scute* gene complex of *Drosophila;* conserved domains in a subset of genes required for neurogenesis and their homology to *c-myc. Cell,* **50,** 415–424.

50. Rijsewijk,F., Schuermann,M., Wagenaar,E., Parren,P., Weigel,D. and Nusse,R. (1987) The *Drosophila* homologue of the mouse mammary oncogen *int-1* is identical to the segment polarity gene *wingless. Cell,* **50,** 649–657.

51. Dickson,C. and Peters,G. (1987) Potential oncogene products related to growth factors. *Nature,* **326,** 833–835.

52. Mercola,M. and Stiles,C.D. (1988) Growth factor superfamilies and mammalian embryogenesis. *Development,* **102,** 451–460.

53. Downward,J., Yarden,Y., Mayes,E., Scrace,G., Tatty,P., Ullrich,A., Schlessinger,J. and Waterfield,M.D. (1984) Close similarity of epidermal growth factor receptor and *v-erb-B* protein sequences. *Nature,* **307,** 521–527.

54. Adamson,E.D. and Meek,J. (1984) The ontogeny of epidermal growth factor during mouse development. *Dev. Biol.,* **103,** 62–71.

55. Padgett,R.W., St Johnson,R.D. and Gelbart,W.M. (1987) A transcript from a *Drosophila* pattern gene predicts a protein homologous to the transforming growth factor β gene family. *Nature,* **325,** 81–84.

56. Wharton,K.A., Johansen,K.M., Xu,T. and Artavonis-Tsakonis,S. (1985) Nucleotide sequence from the neurogenic locus *Notch* implies a gene product that shows homology with proteins containing EGF-like repeats. *Cell,* **43,** 567–581.

57. Greenwald,I. (1985) *lin-12,* a nematode homeotic gene is homologous to a set of mammalian proteins that include epidermal growth factor. *Cell,* **43,** 583–590.

58. Slack,J.M.W., Darlington,B.G., Heath,J.K. and Godsave,S.F. (1987) Mesoderm induction in early *Xenopus* embryos by heparin-binding growth factors. *Nature,* **326,** 197–200.

59. Kimelman,D. and Kirschner,M. (1987) Synergistic induction of mesoderm by FGF and TGF-β and the identification of an mRNA coding for FGF in the early *Xenopus* embryo. *Cell,* **51,** 869–877.

60. Weeks,D.L. and Melton,D.A. (1987) A maternal mRNA localised to the vegetal hemisphere in *Xenopus* egg codes for a growth factor related to TGF-β. *Cell,* **51,** 861–867.

61. Twardzik,D.R. (1985) Differential expression of transforming growth factor α during prenatal development of the mouse. *Cancer Res.,* **45,** 5413–5416.

62. Heath,J.K. and Shi,W.-K. (1986) Developmentally regulated expression of insulin-like growth factors by differentiated murine teratocarcinomas and extraembryonic mesoderm. *J. Embryol. Exp. Morphol.,* **95,** 193–212.

63. Han,V.K., D'Ercole,A.J. and Lund,P.K. (1987) Cellular localization of somatomedin (insulin-like growth factor) mRNA in the human fetus. *Science,* **236,** 193–197.

64. Heath,J.K. and Isacke,C.M. (1984) PC13 embryonal carcinoma-derived growth factor. *EMBO J.,* **3,** 2957–2962.

65. Taira,M., Yoshida,T., Miyagawa,K., Sakemoto,H., Terada,M. and Sugimura,T. (1987) cDNA sequence of human transforming gene *hst* and identification of the coding sequence required for transforming activity. *Proc. Natl. Acad. Sci. USA,* **84,** 2980–2984.

66. McGinnis,W., Levine,M.S., Hafen,E., Kuriowa,A. and Gehring,W.J. (1984) A conserved DNA sequence in homeotic genes of the *Drosophila Antennapedia* and *bithorax* complexes. *Nature,* **308,** 428–433.

67. Gehring,W.J. (1987) Homeo boxes in the study of development. *Science,* **236,** 1245–1252.

68. Bopp,D., Burri,M., Baumgartner,S., Frigerio,G. and Noll,M. (1986) Conservation of a large protein domain in the segmentation gene *paired* and in functionally related genes of *Drosophila. Cell,* **47,** 1033–1040.

69. Berleth,T., Burri,M., Thoma,G.,., Bopp,D., Richstein,S., Frigero,G., Noll,M. and Nusslein-Volhard,C. (1988) The role of localisation of *bicoid* RNA in organising the anterior pattern of the *Drosophila* embryo. *EMBO J.,* **7,** 1749–1756.

70. Saint,R., Kalionis,B., Lockett,T.J. and Eluzur,A. (1988) Pattern formation in the developing eye of *Drosophila melanogaster* is regulated by the homeo-box gene *rough. Nature,* **334,** 151–154.

71. Shepherd,J.C.W., McGinnis,W., Carrasco,A.E., DeRobertis,E.M. and Gehring,W.J. (1984) Fly and frog homeo domains show homologies with yeast mating type regulatory proteins. *Nature,* **310**, 70–71.
72. Burglin,T. (1988) The yeast regulatory gene *PHO2* encodes a homeobox. *Cell,* **53**, 339–340.
73. Way,J.C. and Chalfie,M. (1988) *mec-3*, a homeobox containing gene that specifies differentiation of the touch receptor neurons in *C.elegans. Cell,* **54**, 5–16.
74. Desplan,C., Theis,J. and O'Farrell,P.H. (1988) The sequence specificicity of homeodomain–DNA interaction. *Cell,* **54**, 1081–1090.
75. Hoey,T. and Levine,M. (1988) Divergent homeobox proteins recognise similar DNA sequences in *Drosophila.* Nature, **332**, 858–861.
76. Laughon,A. and Scott,M.P. (1984) Sequence of a *Drosophila* segmentation gene; protein structural homology with DNA-binding proteins. *Nature,* **310**, 25–31.
77. McGinnis,W., Garber,R.L., Wirz,J., Kuroiwa,A. and Gehring,W.J. (1984) A homologous protein-coding sequence in *Drosophila* homeotic genes and its conservation in other metazoans. *Cell,* **37**, 403–408.
78. Hart,C.P., Fainsod,A. and Ruddle,F.H. (1987) Sequence analysis of the murine *Hox-*2.2, -2.3 and -2.4 homeoboxes; evolutionary and structural comparisons. *Genomics,* **1**, 182–195.
79. Graham,A., Papalopulu,N., Lorimer,J., McVey,J.H., Tuddenham,E.G.D. and Krumlauf,R. (1988) Characterisation of a murine homeobox gene, *Hox-2.6*, related to the *Drosophila Deformed* gene. *Genes Dev.,* **2**, 1424–1438.
80. Martin,G.R., Boncinelli,E., Duboule,D., Gruss,P., Jackson,I., Krumlauf,R., Lonai,P., McGinnis,W., Ruddle,F. and Wolgemuth,D. (1987) Nomenclature for homeobox containing genes. *Nature,* **325**, 21–22.
81. Lonai,P., Arman,E., Czosnek,H., Ruddle,F.H. and Blatt,C. (1987) New murine homeoboxes; structure, chromosomal assignment, and differential expression in adult erythropoiesis. *DNA,* **6**, 409–418.
82. Rubin,M.R., King,W., Toth,L.E., Sawczuk,I.S., Levine,M.S., D'Eustachio,P. and Nguyen-Huu,M.C. (1987) The murine *Hox-1.7* homeobox gene; cloning, chromosomal location and expression. *Mol. Cell. Biol.,* **7**, 3836–3841.
83. Featherstone,M.S., Baron,A., Gaunt,S.J., Mattei,M.-G. and Duboule,D. (1988) *Hox-5.1* defines a homeobox containing gene locus on mouse chromosome 2. *Proc. Natl. Acad. Sci. USA,* **85**, 4760–4764.
84. Baron,A., Featherstone,M.S., Hill,R.E., Hall,A., Gaillot,B. and Duboule,D. (1987) Hox-1.6 a mouse homeobox containing gene member of the HOX1 complex. *EMBO J.,* **6**, 2977–2986.
85. LaRosa,G.L. and Gudas,L.J. (1988) Early retinoic acid-induced F9 teratocarcinoma stem cell gene ERA-1; alternative splicing creates transcripts for a homeobox-containing protein and one lacking the homeo-box. *Mol. Cell. Biol.,* **8**, 3906–3917.
86. Wright,C.V.E., Cho,K.W.Y., Fritz,A., Burghlin,T.R. and DeRobertis,E.M. (1987) A *Xenopus laevis* gene encodes both homeobox-containing and homeobox-less transcripts. *EMBO J.,* **6**, 4083–4094.
87. Simeone,A., Pannese,M., Acampora,D., D'Esposito,M. and Boncinelli,E. (1988) At least three human homeoboxes on chromosome 12 belong to the same transcription unit. *Nucleic Acids Res.,* **16**, 5379–5390.
88. Krumlauf,R., Holland,P.W.H., McVey,J.H. and Hogan,B.L.M. (1987) Developmental and spatial pattern of expression of the mouse homeobox gene *Hox-2.1. Development,* **99**, 603–617.
89. Odenwald,W.F., Taylor,C.F., Palmer-Hill,F.J., Friedrich,V., Tani,M. and Lazzarini, R.A. (1987) Expression of a homeo-domain protein in non-contact-inhibited cultured cells and post-mitotic neurons. *Genes Dev.,* **1**, 482–496.
90. Joyner,A.L. and Martin,G.R. (1987) *En-1* and *En-2*, two mouse genes with sequence homology to the *Drosophila engrailed* gene; expression during development. *Genes Dev.,* **1**, 29–38.
91. Hill,R.E., Jones,P.F., Rees,A.R., Sime,C.M., Justice,M.J., Copeland,N.G., Jenkins, N.A., Graham,E. and Davidson,D.R. (1989) A new family of mouse homeobox containing genes; molecular structure, chromosomal location and developmental expression of *Hox-7.1. Genes Dev.,* **3**, 26–37.
92. Gehring,W.J. (1987) The homeobox; structural and evolutionary aspects. In *Molecular Approaches to Developmental Biology.* Alan R.Liss, New York, pp. 115–129.

93. Holland,P.W.H. and Hogan,B.L.M. (1988) Expression of homeobox genes during mouse development; a review. *Genes Dev.*, **2**, 773–782.
94. Jackson,I.J., Schofield,P. and Hogan,B.L.M. (1985) A mouse homeobox gene is expressed during embryogenesis and in adult kidney. *Nature*, **317**, 745–747.
95. Gaunt,S.J. (1988) Mouse homeobox gene transcripts occupy different but overlapping domains in the embryonic germ layers and organs; a comparison of *Hox-3.1* and *Hox-1.5*. *Development*, **103**, 131–144.
96. Sharpe,P.T., Miller,J.R., Evans,E.P., Burtenshaw,M.D. and Gaunt,S.J. (1988) Isolation and expression of a new homeobox gene. *Development*, **102**, 397–407.
97. Holland,P.W.H. and Hogan,B.L.M. (1988) Spatially restricted pattern of expression of the homeobox-containing gene *Hox-2.1* during mouse embryogenesis. *Development*, **102**, 159–174.
98. Dony,C. and Gruss,P. (1987) Specific expression of the *Hox-1.3* homeobox-containing gene in murine embryonic structures originating from or induced by the mesoderm. *EMBO J.*, **6**, 2965–2975.
99. Breier,G., Dressler,G.R. and Gruss,P. (1988) Primary structure and developmental expression of the murine *Hox-3* homeobox gene cluster. EMBO J., **7**, 1329–1336.
100. Davis,C.S., Rossant,J. and Joyner,A.L. (1988) Expression of the homeobox containing gene *En-2* delineates a specific region of the developing mouse brain. *Genes Dev.*, **2**, 361–371.
101. Davidson,D., Graham,E., Sime,C. and Hill,R.E. (1988) A gene with sequence similarity to *Drosophila engrailed* is expressed during the development of the neural tube and vertebrae in the mouse. *Development*, **104**, 305–316.
102. Doe,C.Q., Hiromi,Y., Gehring,W.J. and Goodman,C.S. (1988) Expression and function of the segmentation gene *fushi-tarazu* during *Drosophila* neurogenesis. *Science*, **239**, 170–175.
103. Fritz,A. and DeRobertis,E.M. (1988) *Xenopus* homeobox containing cDNAs expressed in early development. *Nucleic Acids Res.*, **16**, 1453–1469.
104. Harvey,R.P., Tabin,C.J. and Melton,D.A. (1986) Embryonic expression and nuclear localisation of *Xenopus* homeobox (Xhox) gene products. *EMBO J.*, **5**, 1237–1244.
105. Deutsch,U., Dressler,G.R. and Gruss,P. (1988) *Pax-1*, a member of a paired box homologous murine gene family, is expressed in segmented structures during development. *Cell*, **53**, 617–625.
106. Rosenberg,U.B., Schroder,C., Preiss,A., Kienlin,A., Cote,S., Riede,E. and Jackle,H. (1986) Structural homology of the product of the *Drosophila Kruppel* gene with *Xenopus* transcription factor IIIA. *Nature*, **319**, 336–339.
107. Tautz,D., Lehman,R., Schnurch,H., Schuh,R., Seifert,E., Kienlin,A., Jones,K. and Jackle,H. (1987) Finger protein of novel structure encoded by *hunchback*, a second member of the gap class of *Drosophila* segmentation genes. *Nature*, **327**, 383–389.
108. Miller,J., McLachlan,A.D. and Klug,A. (1985) Repetitive zinc-binding domain in the protein transcription factor IIIA from *Xenopus* oocytes. *EMBO J.*, **4**, 1609–1614.
109. Evans,R.M. and Hallenberg,S.M. (1988) Zinc fingers; gilt by association. *Cell*, **52**, 1–3.
110. Chowdhury,K., Deutsch,U. and Gruss,P. (1987) A multigene family encoding several 'finger' structures is present and differentially active in mammalian genomes. *Cell*, **48**, 771–778.
111. Chowdhury,K., Dressler,G.R., Brier,G., Deutsch,U. and Gruss,P. (1988) The primary structure of the murine multifinger gene mKr2 and its specific expression in developing and adult neurons. *EMBO J.*, **7**, 1345–1353.
112. Snow,M.H.L. and Tam,P.P.L. (1979) Is compensatory growth a complicating factor in mouse teratology? *Nature*, **279**, 555–557.
113. Gregg,B.C. and Snow,M.H.L. (1983) Axial abnormalities following disturbed growth in mitomycin C treated mouse embryos. *J. Embryol. Exp. Morphol.*, **73**, 135–149.
114. Falk,D.R. (1977) Genetic mosaics of the *rudimentary* locus of *Drosophila melanogaster*; a genetical investigation of the physiology of pyrimidine synthesis. *Dev. Biol.*, **58**, 134–147.
115. McLaren,A. (1976) Genetics of the early mouse embryo. *Annu. Rev. Genet.*, **10**, 361–388.
116. Patterson,H.F. (1976) *In vivo* and *in vitro* studies on the embryonic lethal oligosyndactylism (*Os*) in the mouse. *J. Embryol. Exp. Morphol.*, **52**, 115–125.
117. Magnuson,T. and Epstein,C.J. (1984) *Oligosyndactyly*; a lethal mutation in the mouse that results in mitotic arrest very early in development. *Cell*, **38**, 823–833.

118. Yee,D., Golden,W., Debrot,S. and Magnuson,T. (1987) Short-term rescue by RNA injection of a mitotic arrest mutation that affects the preimplantation mouse embryo. *Dev. Biol.*, **122**, 256 – 261.

119. Greenspan,R. and O'Brien,M.C. (1986) Genetic analysis of mutations at the *Fused* locus in the mouse. *Proc. Natl. Acad. Sci. USA*, **83**, 4413 – 4417.

120. Bennett,D. (1975) The *T*-locus of the mouse., *Cell*, **6**, 441 – 454.

121. Balling,R., Deutsch,U. and Gruss,P. (1988) *undulated*, a mutation affecting the development of the mouse skeleton has a point mutation in the paired box of *Pax-1*. *Cell*, **55**, 531 – 555.

122. Russel,L.B., Russel,W.L. and Kelly,E.M. (1979) Analysis of the *albino* locus region of the mouse. I. Origin and viability. *Genetics*, **91**, 127 – 139.

123. Russel,L.B., Montgomery,C.S. and Raymer,G.D. (1982) Analysis of the *albino* locus region of the mouse. IV. Characterization of 34 deficiencies. *Genetics*, **100**, 427 – 453.

124. Niswander,L., Yee,D., Rinchik,E.M., Russel,L.B. and Magnuson,T. (1988) The *albino*-deletion complex and early post-implantation survival in the mouse. *Development*, **102**, 45 – 53.

125. Lewis.S. (1978) Developmental analysis of the lethal effects of homozygotes for the c^{25H} deletion in the mouse. *Dev. Biol.*, **65**, 553 – 557.

126. Kwon,B.S., Wakulchik,M., Haq,A.K., Halaban,R. and Kestler,D. (1988) Sequence analysis of mouse tyrosinase cDNA and the effect of melanotropin on its gene expression. *Biochem. Biophys. Res. Commun.*, **153**, 1301 – 1309.

127. Muller,G., Ruppert,S., Schmid,E. and Schutz,G. (1988) Functional analysis of alternatively spliced tyrosinase gene transcripts. *EMBO J.*, **7**, 2723 – 2730.

128. Rinchik,E.M., Russel,L.B., Copeland,N.G. and Jenkins,N.A. (1986) The *dilute-short ear* (*d-se*) complex of the mouse; lessons from a fancy mutation. *Trends Genet.*, **1**, 170 – 176.

129. Russel,L.B. (1971) Definition of functional units in a small chromosomal segment of the mouse and its use in interpreting the nature of radiation induced mutations. *Mutat. Res.*, **11**, 107 – 123.

130. Rinchik,E.M., Russel,L.B., Copeland,N.G. and Jenkins,N.A. (1986) Molecular genetic analysis of the *dilute-short ear* region of the mouse., *Genetics*, **112**, 321 – 342.

131. Jackson,I.J. (1988) A cDNA encoding tyrosinase-related protein maps to the *brown* locus of mouse. *Proc. Natl. Acad. Sci. USA*, **85**, 4392 – 4396.

132. Castle,W.E. and Little,C.C. (1910) On a modified mendelian ratio among yellow mice. *Science*, **32**, 868 – 870.

133. Russel,L.B., McDaniel,M.N.C. and Woodiel,F.N. (1963) Crossing-over within the *a*-locus of the mouse. *Genetics*, **48**, 907a.

134. Siracusa,L.D., Russel,L.B., Eicher,E.M., Corrow,D.J,. Copeland,N.G. and Jenkins,N.A. (1987) Genetic organization of the *agouti*-region of the mouse. *Genetics*, **117**, 93 – 100.

135. Papaioannou,V.E. and Gardner,R.L. (1979) Investigation of the *Lethal yellow* (A^yA^y) embryo using mouse chimaeras. *J. Embryol. Exp. Morphol.*, **42**, 153 – 163.

136. Papaioannou,V.E. and Mardon,H. (1983) *lethal nonagouti* (a^x); description of a second embryonic lethal at the *agouti* locus. *Dev. Genet.*, **4**, 21 – 29.

137. Lovett,M., Cheng,Z.-Y., Lamela,E.M., Yokoi,T. and Epstein,C.J. (1987) Molecular markers for the *agouti* coat color locus of the mouse. *Genetics*, **115**, 747 – 754.

138. Willison,K. (1986) 't-party' at the Jackson Laboratory. *Trends Genet.*, **2**, 305 – 306.

139. Silver,L.M. (1985) Mouse *t*-haplotypes. *Annu. Rev. Genet.*, **19**, 179 – 208.

140. Schimenti,J., Cebra-Thomas,J.H., Decker,C.L., Islam,S.D., Pilder,S.H. and Silver,L.M. (1988) A candidate gene family of the mouse *t-complex responder* (*Tcr*) locus responsible for haploid effects on sperm function. *Cell*, **55**, 71 – 78.

141. Shin,H.-S., Bennett,D. and Artzt,K. (1984) Gene mapping within the *T/t* complex of the mouse. IV. The inverted MHC is intermingled with several *t*-lethal genes. *Cell*, **39**, 573 – 578.

142. Klein,J. and Hammerberg,C. (1977) The control of differentiation by the *T*-complex. *Immunol. Rev.*, **33**, 70 – 104.

143. Shedovsky,A., Guenet,J.-L., Johnson,L.L. and Dove,W.F. (1986) Induction of recessive lethal mutations in the *T/t*-H2 region of the mouse genome by a point mutagen. *Genetical Res.*, **47**, 135 – 142.

144. Abe,K., Wei,J.-F., Wei,F.-S., Hsu,Y.-C., Uehara,H., Artzt,K. and Bennett,D. (1988) Searching for coding sequences in the mammalian genome: the H-2K region of the

mouse is replete with genes expressed in embryos. *EMBO J.*, **7**, 3441–3449.

145. Gridley,T., Soriano,P. and Jaenisch,R. (1987) Insertional mutagenesis in mice. *Trends Genet.*, **3**, 162–166.

146. Woychik,R.P., Stewart,T.A., Davis,L.G., D'Eustachio,P. and Leder,P. (1985) An inherited limb deformity created by insertional mutagenesis in a transgenic mouse. *Nature*, **318**, 36–40.

147. Jenkins,J.A. and Copeland,N.G. (1985) High frequency germline acquisition of ecotropic MuLV proviruses in SWR/J × RF/J hybrid mice. *Cell*, **43**, 811–819.

148. Schnieke,A., Harbers,K. and Jaenisch,R. (1983) Embryonic lethal mutation in mice induced by retrovirus insertion at the α1(I)collagen gene. *Nature*, **304**, 315–320.

149. Lohler,J., Timpl,R. and Jaenisch,R. (1984) Embryonic lethal mutation in mouse collagen I gene causes rupture of blood vessels and is associated with erythropoietic and mesenchymal cell death. *Cell*, **38**, 597–607.

150. McNeish,J.D., Scott,W.J. and Potter,S.S. (1988) Legless, a novel mutation found in PHT1-1 transgenic mice. *Science*, **241**, 837–839.

151. Soriano,P., Gridley,T. and Jaenisch,R. (1987) Retroviruses and insertional mutagenesis; proviral integration at the Mov-34 locus leads to early embryonic death. *Genes Dev.*, **1**, 366–375.

152. Mark,W.H., Signorelli,K. and Lacy,E. (1985) An insertional mutation in a transgenic mouse line results in developmental arrest at day 5 of gestation. *Cold Spring Harbor Symp. Quant. Biol.*, **50**, 453–463.

153. Covarrubias,L., Nishida,Y. and Mintz,B. (1985) Early developmental mutations due to DNA rearrangements in transgenic mouse embryos. *Cold Spring Harbor Symp. Quant. Biol.*, **50**, 447–452.

154. Shani,M. (1986) Tissue-specific and developmental expression of a chimeric actin-globin gene in transgenic mice. *Mol. Cell. Biol.*, **6**, 2624–2631.

155. Covarrubias,L., Nishida,Y. and Mintz,B. (1986) Early post-implantation embryonic lethality due to DNA rearrangements in a transgenic mouse strain. *Proc. Natl. Acad. Sci. USA*, **83**, 6020–6024.

156. Palmiter,R.D. and Brinster,R.L. (1986) Germ-line transformation of mice. *Annu. Rev. Genet.*, **20**, 465–499.

157. Jahner,D. and Jaenisch,R. (1980) Integration of Moloney leukaemia virus into the germline of mice; correlation between site of integration and virus activity. *Nature*, **287**, 456–458.

158. Jaenisch,R. (1980) Retroviruses and embryogenesis; microinjection of Moloney leukaemia virus into midgestation mouse embryos. *Cell*, **19**, 181–188.

159. Groner,B., Schonenberger,C.A. and Andres,A.C. (1987) Targeted expression of the *ras* and *myc* oncogenes in transgenic mice. *Trends Genet.*, **3**, 306–308.

160. Gordon,J. (1986) A foreign dihydrofolate reductase gene in transgenic mice acts as a dominant mutation. *Mol. Cell. Biol.*, **6**, 2158–2167.

161. Walder,J. (1988) Antisense DNA and RNA; progress and prospects. *Genes Dev.*, **2**, 502–504.

162. Bevilacqua,A., Erickson,R.P. and Hieber,V. (1988) Antisense RNA inhibits endogenous gene expression in mouse pre-implantation embryos; lack of double-stranded RNA 'melting' activity. *Proc. Natl. Acad. Sci. USA*, **85**, 831–835.

163. Ao,A., Monk,M., Lovell-Badge,R.H. and Melton,D.W. (1988) Expression of injected HPRT minigene DNA in mouse embryos and its inhibition by antisense DNA. *Development*, **104**, 465–471.

164. Katsuki,M., Sato,M., Kimura,M., Yokoyama,M., Kobayashi,K. and Nomura,T. (1988) Conversion of normal behavior to shiverer by myelin basic protein antisense cDNA in transgenic mice. *Science*, **241**, 593–595.

165. Cullen,B.R. (1986) *Trans*-activation of human immunodeficiency virus occurs via a bimodal mechanism. *Cell*, **46**, 973–982.

166. Kakidara,H. and Ptashne,M. (1988) GAL4 activates gene expression in mammalian cells. *Cell*, **52**, 161–167.

167. Webster,N., Jin,J.-R., Green,S., Hollis,M. and Chambon,P. (1988) The yeast UAS[G] is a transcriptional enhancer in human HeLa cells in the presence of the GAL4 *trans*-activator. *Cell*, **52**, 169–178.

168. O'Kane,C.J. and Gehring,W.J. (1987) Detection *in situ* of genome regulatory elements in *Drosophila*. Proc. Natl. Acad. Sci. USA, **84**, 9123–9127.

169. Allen,N.D., Cran,D.G., Barton,S.C., Hettle,S., Reik,W. and Surani,M.A. (1988) Transgenes as probes for active chromosomal domains in mouse development. *Nature,* **333**, 852–855.
170. Ponder,B.A.J. (1987) Cell marking techniques and their application. In *Mammalian Development: A Practical Approach.* Monk,M. (ed.), IRL Press, Oxford, pp. 115–138.
171. Ponder,B.A.J., Schmidt,G.H., Wilkinson,M.M., Wood,M.J., Monk,M. and Reid,A. (1985) Derivation of mouse intestinal crypts from single progenitor cells. *Nature,* **313**, 689–691.
172. Price,J. (1987) Retroviruses and the study of cell lineage. *Development,* **101**, 409–419.
173. Soriano,P. and Jaenisch,R. (1986) Retroviruses as probes for mammalian development; allocation of cells to the somatic and germ cell lineages. *Cell,* **46**, 19–29.
174. Rossant,J. (1986) Retroviral mosaics; a new approach to cell lineage analysis in the mouse embryo. *Trends Genet.,* **2**, 302–303.
175. Sanes,J.R., Rubenstein,J.L.R. and Nicolas,J.-F. (1986) Use of a recombinant retrovirus to study postimplantation cell lineage in the mouse embryo. *EMBO J.,* **5**, 3133–3142.
176. Beddington,R.S.P. (1988) Toxigenics; strategic cell death in the embryo. *Trends Genet.,* **4**, 1–2.
177. Palmiter,R.D., Behringer,R.R., Quaife,C.J., Maxwell,F., Maxwell,I.A. and Brinster, R.L. (1987) Cell lineage ablation in transgenic mice by cell-specific expression of a toxin gene. *Cell,* **50**, 435–443.
178. Breitman,M.L., Clapoff,S., Rossant,J., Tsui,L.-C., Glode,L.M., Maxwell,I.H. and Bernstein,A. (1987) Genetic ablation; targeted expression of a toxin gene causes microphthalmia in transgenic mice. *Science,* **238**, 1563–1565.
179. Behringer,R.R., Mathews.L.S., Palmiter,R.D. and Brinster,R.L. (1988) Dwarf mice produced by genetic ablation of growth hormone expressing cells. *Genes Dev.,* **2**, 453–461.
180. Surani,M.A.H., Barton,S.C. and Norris,M.L. (1984) Development of reconstituted mouse eggs suggests imprinting of the genome during gametogenesis. *Nature,* **308**, 548–550.
181. McGrath,J. and Solter,D. (1984) Completion of mouse embryogenesis requires both the maternal and paternal genomes. *Cell,* **37**, 179–183.
182. Solter,D. (1987) Inertia of the embryonic genome in mammals. *Trends Genet.,* **3**, 23–27.
183. Cattanach,B.M. and Kirk,M. (1985) Differential activity of maternally and paternally derived chromosomal regions in mice. *Nature,* **315**, 496–498.
184. Lyon,M.F. (1983) The use of robertsonian translocations for the study of nondisjunction. In *Radiation Induced Chromsomal Damage in Man.* Ishihara,T. (ed.), Alan Liss, New York, pp. 327–346.
185. Johnson,D.R. (1974) *Hairpin-tail;* a case of post-reductional gene action in the mouse egg. *Genetics,* **76**, 795–805.
186. Winking,H. and Silver,L. (1984) Characterisation of a recombinant mouse haplotype that expresses a dominant lethal maternal effect. *Genetics,* **108**, 1013–1026.
187. McGrath,J. and Solter.D. (1984) Maternal T^{hp} lethality in the mouse is a nuclear, not cytoplasmic, defect. *Nature,* **308**, 550–551.
188. Surani,M.A., Reik,W. and Allen,N.D. (1988) Transgenes as molecular probes for genomic imprinting. *Trends Genet.,* **4**, 59–62.
189. Beechey,C., Cattanach,B.M. and Searle,A.G. (1988) Genetic imprinting map. *Mouse News Lett.,* **80**, 48–49.
190. Reik,W., Collick,A., Norris,M.L., Barton,S.C. and Surani,M.A.H. (1987) Genomic imprinting determines methylation of parental alleles in transgenic mice. *Nature,* **328**, 248–251.
191. Sapienza,C., Peterson,A.C., Rossant,J. and Balling,R. (1987) Degree of methylation of transgenes is dependent on gamete of origin. *Nature,* **328**, 251–254.
192. Swain,J.L., Stewart,T.A. and Leder,P. (1987) Parental legacy determines methylation and expression of an autosomal transgene; a molecular mechanism for parental imprinting. *Cell,* **50**, 719–727.
193. Hadchouel,M., Farza,H., Simon,D., Tiollais,P. and Pourcel,C. (1987) Maternal inhibition of hepatitis B surface antigen gene expression in transgenic mice correlates with *de novo* methylation. *Nature,* **329**, 454–456.
194. Sanford,J., Forrester,L., Chapman,V., Chandley,A. and Hastie,N.D. (1984) Methylation

patterns of repetitive DNA sequences in germ cells of *Mus musculus. Nucleic Acids Res.,* **12**, 2823–2836.

195. Monk,M., Boubelik,M. and Lehnert,S. (1987) Temporal and regional changes in DNA methylation in the embryonic, extraembryonic and germ cell lineages during mouse embryo development. *Development,* **99**, 371–382.

196. Andrews,P.W., Oosterhuis,J.W. and Damjanov,I. (1987) Cell lines from human germ cell tumours. In *Teratocarcinomas and Embryonic Stem Cells: A Practical Approach.* Robertson,E.J. (ed.), IRL Press, Oxford, pp. 207–248.

197. Martin,G.R. (1980) Teratocarcinomas and mammalian embryogenesis. *Science,* **209**, 768–776.

198. Damjanov,I., Damjanov,A. and Solter,D. (1987) Production of tumours from embryos transplanted to ectopic sites. In *Teratocarcinomas and Embryonic Stem Cells: A Practical Approach.* Robertson,E.J. (ed.), IRL Press, Oxford, pp. 1–18.

199. Stewart,T.A. and Mintz,B. (1981) Successive generations of mice produced from an established cultured line of euploid teratocarcinoma cells. *Proc. Natl. Acad. Sci. USA,* **78**, 6314–6318.

200. Evans,M.J. and Kaufman,M.H. (1981) Establishment in culture of pluripotential cells from mouse embryos. *Nature,* **292**, 154–156.

201. Martin,G.R. (1981) Isolation of a pluripotent cell line from early mouse embryos cultured in medium conditioned by teratocarcinoma stem cells. *Proc. Natl. Acad. Sci. USA,* **78**, 7634–7638.

202. Robertson,E.J. (1987) Embryo-derived stem cells. In *Teratocarcinomas and Embryonic Stem Cells: A Practical Approach.* Robertson,E.J. (ed.), IRL Press, Oxford, pp. 71–112.

203. Smith,A.G. and Hooper,M.L. (1987) Buffalo rat liver cells produce a diffusible activity which inhibits differentiation of murine embryonic carcinoma and embryonic stem cells. *Dev. Biol.,* **121**, 1–9.

204. Evans,M.J. and Kaufman,M.H. (1983) Pluripotent cells grown directly from normal mouse embryos. *Cancer Surv.,* **2**, 185–207.

205. Kuehn,M.R., Bradley,A., Robertson,E.J. and Evans,M.J. (1987) A potential animal model for Lesch–Nyhan syndrome through introduction of HPRT mutations into mice. *Nature,* **326**, 295–298.

206. Magnuson,T., Epstein,C.J., Silver,L. and Martin,G.R. (1982) Pluripotent embryonic stem cell lines can be derived from t^{w5}/t^{w5} blastocysts. *Nature,* **298**, 750–753.

207. Bradley,A. (1987) Production and analysis of chimaeric mice. In *Teratocarcinomas and Embryonic Stem Cells: A Practical Approach.* Robertson,E.J. (ed.), IRL Press, Oxford, pp. 113–151.

208. Jackson,I.J. (1985) Genetics and biology of mouse melanocytes; mutation, migration and interaction. *Trends Genet.,* **1**, 321–326.

209. Lovell-Badge,R.H. (1987) Introduction of DNA into embryonic stem cells. In *Teratocarcinomas and Embryonic Stem Cells: A Practical Approach.* Robertson,E.J. (ed.), IRL Press, Oxford, pp. 153–182.

210. Lovell-Badge,R.H., Bygrave,A., Bradley,A., Robertson,E.J., Tilly,R. and Cheah,K.S.E. (1987) Tissue-specific expression of the human type II collagen gene in mice. *Proc. Natl. Acad. Sci. USA,* **84**, 2803–2807.

211. Hooper,M.L. (1987) Isolation of genetic variants and fusion hybrids from embryonic carcinoma cells. In *Teratocarcinomas and Embryonic Stem Cells: A Practical Approach.* Robertson,E.J. (ed.), IRL Press, Oxford, pp. 51–70.

212. Robertson,E.J., Bradley,A., Kuehn,M.R. and Evans,M.J. (1987) Germline transmission of genes introduced into cultured pluripotential cells by retroviral vectors. *Nature,* **323**, 445–448.

213. Taketo,M. and Tanaka,M. (1987) A cellular enhancer of retroviral gene expression in embryonic carcinoma cells. *Proc. Natl. Acad. Sci. USA,* **84**, 3748–3752.

214. Hooper,M.L., Hardy,K., Handyside,A., Hunter,S. and Monk,M. (1987) HPRT-deficient (Lesch–Nyhan) mouse embryos derived from germline colonisation by cultured cells. *Nature,* **326**, 292–295.

215. Orr-Weaver,T.L., Szostak,J.W. and Rothstein,R.J. (1981) Yeast transformation; a model system for the study of recombination. *Proc. Natl. Acad. Sci. USA,* **78**, 6354–6358.

216. Smithies,O., Gregg,R.G., Boggs,S.S., Koralewski,M.A. and Kucherlapati,R.S. (1985) Insertion of DNA sequences into the human chromosomal β-globin locus by homologous recombination. *Nature,* **317**, 230–234.

217. Thomas,K.R., Folger,K.R. and Cappechi,M.R. (1986) High frequency targeting of genes to specific sites in the mammalian genome. *Cell,* **44**, 419–428.

218. Thomas,K.R. and Cappechi,M.R. (1987) Site-directed mutagenesis by gene targeting in mouse embryo-derived stem cells. *Cell,* **51**, 503–512.

219. Doetschman,T., Gregg,R.G., Maeda,N., Hooper,M.L., Melton,D.W., Thompson,S. and Smithies,O. (1987) Targeted correction of a mutant HPRT gene in mouse embryonic stem cells. *Nature,* **330**, 576–578.

Index